An Introduction to
Real Analysis

An Introduction to
Real Analysis

V.K. Bhat

Alpha Science International Ltd.
Oxford, U.K.

V.K. Bhat
School of Mathematics
SMVD University
P.O. SMVD University
Katra 182320 (J&K)

ALPHA SCIENCE INTERNATIONAL LTD.
7200 The Quorum, Oxford Business Park North
Garsington Road, Oxford OX4 2JZ, U.K.

www.alphasci.com

Printed from the camera-ready copy provided by the Author.

ISBN 978-1-84265-705-8

Printed in India

Dedicated to
My beloved wife Late Sunita

PREFACE

The philosophy of Understanding Analysis is to focus attention on questions that give analysis its inherent fascination. Does the Cantor set contain any irrational numbers? Can the set of points where a function is discontinuous be arbitrary. Are derivatives continuous? Are derivatives integrable? Is an indefinitely differentiable function necessarily the limit of its Taylor series? In giving these topics center stage, the hard work of a rigorous study is justified by the fact that they are inaccessible without it.

This book has been designed for under graduate students in mathematics and statistics. It presupposes a general background in undergraduate mathematics and specific acquaintance with the material in an undergraduate course on the fundamental concepts of analysis. I have attempted to cover the basic material that every graduate should know in the classical theory of functions pr real variable. The treatment of material given here is quite standard in graduate courses of this sort.

The book starts with the essential properties of rational and irrational numbers and using Dedekind's axiom, the properties of real numbers are established. Although some fairly sophisticated topics are brought in early to advert and motivate the upcoming material, the main body of each chapter consists of a lean and focused treatment of the core topics that make up the center of most courses in analysis. Fundamental results about completeness, compactness, sequential and functional limits, continuity differentiation, and integration are all incorporated. In the chapter on integration, for instance, the exposition revolves around deciphering the relationship between continuity and the Riemann integral. Enough properties of the integral are obtained to justify a proof of the Fundamental Theorem of Calculus, but the theme of the chapter is the pursuit of a characterization of integrable functions in terms of continuity. In the case of integration, this point is made explicitly by including some relatively recent developments on the generalized Riemann integral in the topics of the last chapters.

The topics in the first, second, fourth and sixth chapters of this volume are related to one another in many ways. The material of this book is given with more details and with many examples and motivation that is usually done. Special effort has been made to include all necessary details. Each chapter is provided with a set of exercises which have straightforward proofs.

The purpose of this book has been to provide a development of the subject-matter which is well motivated, rigorous and at the same time not too pedantic. Efforts have also been taken to give proofs of number of theorems in a some what modified form. I shall feel greatly rewarded if the users of this book and it really beneficial and friendly. Some errors are unavoidable in this work. I shall be grateful to the readers for bringing the errors to my notice.

V.K. Bhat

CONTENTS

INTRODUCTION

Understating of Mathematics depend on the knowledge of number. In fact a major part of mathematics bases its development on numbers and their multifarious properties.

Mathematical analysis deals with study of structure of mathematics that is why it is often called Grammar of mathematics. The basic tools for such a study are provided by real numbers. Consequently, Real analysis forms the foundation of mathematical Analysis.

In our modern approach we start directly from real numbers and then pass on the related concepts. Though there are many types of numbers, the justification to sudden reference to real numbers lies in the fact that calculus we are going to study now is calculus of real numbers.

Another point worth mentioning in this regard is that Cantor's work not only revolutionized the classical ideas but gave a new language to Mathematics, namely Language of set Theory which has made simple many results of Mathematics. In the development of calculus too, set theory is now a day an indispensable tool.

Sequences and series from an important component of Mathematical Analysis and first rigorous treatment of sequences and Series was made by George Cantor (1845-1918) and A. Cauchy (1789-1857).

It has been provided by them that sequences and their convergence are essentially most effective tools of Mathematics. Series is such an important mathematical unit that a lot of Mathematical Analysis has been developed on this and a sizable amount of present day research centers around series.

We are familiar with various operations like addition, subtraction etc. here, we introduce a typical operation called Limit Function performable on functions. Limit operations play a fundamental role in calculus and most of the concepts here requires this operation.

The motion of differentiability is as basic that of continuity but more useful than that. Both these play important role in entire calculus. Both Sir Isaac Newton (1642-1727) and G.W. Leibohite (1646-1716) are accredited for independent discovery of these notions derivability and differentiability.

Classes of theorems in analysis are referred as Mean Value Theorems or Law of Men because of their implication as average value of function over an interval.

At elementary stage the subject of Integration is generally introduced as the inverse of differentiation, so that a function F is called an integral of a function f if $F'(x) = f(x)$, for all values of x belonging to the domain of function. Realization that the subject could be looked upon as inverse of differentiation came afterwards. The reference to integration from summation point of view was always associated with the geometrical concepts.

The first approach quite naturally started based on intuitive ideas of a sum and in effect as the limit of sum, now-a-days known as Riemann sum.

To formulate an independent theory of integration, the German mathematician, G.F.B Riemann (1826 - 1866), gave a purely arithmetic treatment to the subject and thus developed a subject entirely free from the intuitive dependence on geometrical concepts and achieved a remarkable success and is known as Riemannian theory of Integration.

This theory plays a fundamental role in analysis. The last and the present century saw many generalizations of this theory to enlarge the scope of integration for larger classes of functions.

Chapter1

NUMBER SYSTEM

Numbers are Mathematical quantities with wonderful properties. Mathematical analysis study concepts are related in some way to real numbers so we begin our study of analysis with real number system.

The system of real numbers is evolved as a result of process of successive extensions of system of natural numbers. Rational numbers and irrational numbers together constitute what is known as system of real numbers.

We assume with the existence of positive integers
1, 2, 3, ...
These are also called as natural numbers or counting numbers.On including 0 to this set we get non negative integers. Now consider the equation $x + p = 0$, where $p = 1, 2, 3, ...$ The set of solution of this equation is $x = -p$. This gives rise to the numbers -1, -2, -3, ... The set of these numbers is called the set of negative integers. $\mathbb{Z} = \{..., -3, -2, -1, 0, 1, 2, 3, ...\}$ is called the set of integers.

We now establish a correspondence between the elements of the set $\mathbb{R}$ and the points on a line. This line is called as number line. Consider the equation $qx - p = 0$, where p and q are integers and $q \neq 0$. Then the solution of this equation is $x = \frac{p}{q}$. Such numbers of the form $\frac{p}{q}$, where p and q are integers and $q \neq 0$ are called rational numbers. The set of rational numbers is denoted by $\mathbb{Q}$. To be rational, a number ought to have at least one fractional representation. For example, the number $\left(\frac{\sqrt{7}+1}{2}\right)^2 + \left(\frac{\sqrt{7}-1}{2}\right)^2$ may not at first look rational, but it simplifies to 24 which is $24 = \frac{24}{1}$ a rational fraction.

On the other hand, the number $\sqrt{2}$ by itself is not rational and is called irrational. This is by no means a definition of irrational numbers. In Mathematics, it's not quite true that what is not rational is irrational. Irrationality is a term reserved for a very special kind of numbers. However, there are numbers which are neither rational nor irrational (for example, infinitesimal numbers are neither rational nor irrational). The set $\mathbb{Q} = \{\frac{p}{q} \mid p$ and q are integers and $q \neq 0\}$. These numbers fill the gaps between the integers as can be seen on the number line. Thus every element in the set $\mathbb{Q}$ corresponds to some point on the number line but the converse is not true; e.g there exists a point on the number line which corresponds to $\sqrt{2}$ but $\sqrt{2}$ does not belong the set $\mathbb{Q}$. Such type of numbers are called as irrational numbers. The union of rational and irrational numbers forms the set of real numbers and this set is denoted by $\mathbb{R}$. Thus $\mathbb{R} = \{x \mid x$ is a rational or an irrational number$\}$.

Remark 1.1.   (1) Suppose a and b are integers. Then $a+b$, $a-b$, and $a \times b$ are also integers.

(2) Suppose a and b are rational numbers, then $a + b$, $a - b$, $a \times b$ and $\frac{a}{b}$ (where $b \neq 0$) are all rational numbers.

Thus the set of integers is closed with respect to operations $+$, $-$ and $\times$.

Remark 1.2. If a is a rational number and b an irrational number, then $a + b$, $a - b$ and $a \times b$ and $\frac{a}{b}$(where $b \neq 0$) are all irrational numbers.

Remark 1.3. Irrational numbers are not closed under addition, subtraction, multiplication or division. Let $a = \sqrt{2}$ -1 and $b = 3 - \sqrt{2}$. Then $a + b = (\sqrt{2} - 1) + (3 - \sqrt{2}) = 2$, which is a rational number. Hence irrational numbers are not closed under addition.

Similarly irrational numbers are not closed under subtraction, multiplication or division.

Exercise 1.4. Prove that $\sqrt{2}$ is an irrational number.

Proof. If possible suppose $\sqrt{2}$ is a rational number. So, let $\sqrt{2} = \frac{p}{q}$, where p and q are integers and $q \neq 0$. With out loss of generality p, q are co prime to each other.

Therefore,

$$p^2 = 2q^2 \tag{1}$$

Which implies that p^2 is an even number. So, p must be even and let $p = 2m$, where m is an integer. From (1), we get that $(2m)^2 = 2q^2$, which implies that $4m^2 = 2q^2$ or $q^2 = 2m^2$. So q^2 is even, i.e. q is even. Therefore, p and q are both even. But p and q are co prime, and we get a contradiction. Hence $\sqrt{2}$ is an irrational number. $\qquad\square$

Theorem 1.5. (Gauss's Theorem) *Consider a polynomial equation of degree n, i.e $a_0x^n + a_1x^{n-1} + a_2x^{n-2} + ... + a_{n-1}x + a_n = 0$, where a_i is an integer for all $i = 1, 2, 3, ..., n$; $a_0 \neq 0$ and n is a positive integer. Suppose this equation has a rational root $\frac{p}{q}$, where p and q are co prime, then q must divide a_0 and p must divide a_n.*

Proof. Given a polynomial equation of degree n; $a_0x^n + a_1x^{n-1} + a_2x^{n-2} + ... + a_{n-1}x + a_n = 0$, where a_i are integers for all $i = 1, 2, 3, ..., n$; $a_0 \neq 0$, and n is a positive integer. Also given that this equation has a rational root $\frac{p}{q}$, where p and q are co prime. Therefore,

$$a_0\left(\frac{p^n}{q^n}\right) + a_1\left(\frac{p^{n-1}}{q^{n-1}}\right) + a_2\left(\frac{p^{n-2}}{q^{n-2}}\right) + ... + a_{n-1}\left(\frac{p}{q}\right) + a_n = 0$$

i.e.

$$a_0\left(\frac{p^n}{q^n}\right) = -\left[a_1\left(\frac{p^{n-1}}{q^{n-1}}\right) + a_2\left(\frac{p^{n-2}}{q^{n-2}}\right) + ... + a_{n-1}\left(\frac{p}{q}\right) + a_n\right]$$

or,

$$a_0\left(\frac{p^n}{q}\right) = -\left[a_1(p^{n-1}) + a_2(p^{n-2}/q) + ... + a_{n-1}pq^{n-2} + a_nq^{n-1}\right]$$

The R.H.S of this equation is clearly an integer which implies $a_0\left(\frac{p^n}{q}\right)$ must also be an integer. Since p and q are co prime, so q must divide a_0.

Similarly consider the reciprocal equation;

$$a_nx^n + a_{n-1}x^{n-1} + a_{n-2}x^{n-2} + ... + a_1x + a_0 = 0$$

where a_i is an integer for all $i = 1, 2, 3, ..., n$; $a_n \neq 0$ and n is a positive integer. Its root will be $\frac{q}{p}$, and it can be proved that p divides a_n. Hence the proof.   □

Using above Theorem (1.12), we can easily prove that $\sqrt{2}$ is an irrational number.

Exercise 1.6. Prove that $\sqrt{2}$ is an irrational number.

Proof. Consider the equation $x^2 - 2 = 0$. Here $a_0 = 1$, $a_1 = 0$ and $a_2 = -2$. Suppose this equation has a rational root $\frac{p}{q}$, where p and q are co prime. Then by Gauss's Theorem (1.12), q divides 1 and p divides -2. This implies $q = -1, 1$ and $p = -1, 1, -2, 2$. Therefore, $\frac{p}{q} = -1, 1, -2, 2$. But none of these numbers satisfies the equation $x^2 - 2 = 0$. Hence it follows that $\sqrt{2}$ is an irrational number.   □

Exercise 1.7. Prove that $\sqrt{p}$ is an irrational number.

Proof. Consider the equation $x^2 - p = 0$, where p is prime. Here $a_0 = 1$, $a_1 = 0$ and $a_2 = -p$. If possible suppose that this equation has a rational root $\frac{r}{s}$, where r and s are co prime. Then by Gauss's Theorem (1.12), s divides 1 and r divides $-p$. This implies $s = -1, 1$ and $r = -1, 1, -p, p$. Therefore, $r/s = -1, 1, -p, p$. But none of these numbers satisfies the equation $x^2 - p = 0$. Hence it follows that $\sqrt{p}$ is an irrational number.   □

Exercise 1.8. Prove that $\sqrt{\frac{2}{3}}$ is an irrational number.

Proof. Consider the equation $3x^2 - 2 = 0$. Here $a_0 = 3$, $a_1 = 0$ and $a_2 = -2$. If possible suppose that this equation has a rational root $\frac{p}{q}$, where p and q are co prime. Then by Gauss's Theorem (1.12), q divides 3 and p divides -2. This implies $q = -1, 1, -3, 3$ and $p = -1, 1, -2, 2$. Therefore, $\frac{p}{q} = -1, 1, \frac{-1}{3}, \frac{1}{3}, -2, 2, \frac{2}{3}, \frac{2}{3}$. But none of these numbers satisfies the equation $3x^2 - 2 = 0$. Hence it follows that $\sqrt{\frac{2}{3}}$ is an irrational number.   □

Bounded Sets

Definition 1.9. Let E be a set of real numbers. Then E is said to be bounded above or bounded on the right if there exists a real number b such that $x \leq b$, for all $x \in$ E. The element b is called the upper bound of set E. In other words it follows from the definition that a set E is bounded above by the element b if no element of the set E lies on the right of b.

A set T of real numbers is said to be bounded below or bounded on the left if there exists a real number a such that $x \geq a$, for all $x \in T$. The element a is called the lower bound of set T. In other words a set T is bounded below by the element a if no element of the set T lies on the left of a.

A set K of real numbers is said to be bounded if it is bounded both above and below.

For example consider the set of positive integers $\mathbb{N} = \{1, 2, 3, ...\}$. This set is bounded below by 1 because no element of this set lies to the left of 1. Also the set of negative integers $\{..., -3, -2, -1\}$ is bounded above by -1 because no element of this set lies to the right of -1. Consider the set $A = \{1, 2, 3, ..., 100\}$. This set is bounded above by 100 and below by 1. So, A is a bounded set. Obviously every finite set is a bounded set.

Definition 1.10. An element l is said to be the least upper bound (LUB)or Supremum of a set E if it satisfies following two properties:

(1) $x \leq l$ for all $x \in E$

(2) Given a positive number ϵ (how so ever small), $l - \epsilon < x$ for at least one $x \in E$.

We write $Sup(E) = l$.

An element g is said to be the greatest lower bound (GLB) or Infimum of a set E if it satisfies following two properties:

(1) $x \geq g$, for all $x \in E$

(2) Given a positive number ϵ (how so ever small), $g + \epsilon > x$ for at least one $x \in E$.

We write $Inf(E) = l$.

$*$ **Dedekind's Axiom:** Suppose a set of real numbers E is divided into two non empty subsets L and R such that $l \leq r$ for all $l \in L$, $r \in R$. Then there is a dividing number η such that every number less than η is in L and every number greater than η is in R, where $\eta \in L$ or $\eta \in R$. If $\eta \in L$, then η is called greatest element of L and if $\eta \in R$, then η is called least element of R.

Theorem 1.11. (Dedekind's Theorem) *If S is a non empty set which is bounded above, then the least upper bound of S exists.*

Proof. Suppose S is a non empty set. Divided S into two non empty subsets L and R such that

$$x \in L, \text{ if } x < s \text{ for at least one } s \in S$$

and

$$x \in R, \text{ if } x \geq s \text{ for all } s \in S.$$

Then $x \in L$ or $x \in R$. Both L and R are nonempty because, S is non empty, so $s \in S$ and obviously $x = s - 1 \in L$ (because $x < s$). Since S is bounded above, so there exists a real number k such that $x \leq k$, for all $x \in S$ or $k \geq x$ for all $x \in S$. This implies that $k \in R$. Let $l \in L$. Then by definition of L, for at least one $s \in S$,

$$l < s \tag{2}$$

Also if $r \in R$, then for above $s \in S$,

$$r \geq s \tag{3}$$

From (2) and (3), we get that $l < s \leq r$, which implies that $l < r$. Therefore it follows that every element of L is less than every element of R. So by Dedekind's axiom ($*$ above) it follows that there exists a dividing factor ξ such that $\xi - \epsilon \in L$, where ϵ is an arbitrary small positive number and $\xi + \epsilon \in R$, where $\epsilon \in L$ or $\epsilon \in R$.

If possible suppose $\epsilon \notin R$, then $\epsilon \in L$. So, by the definition of L; for at least one $s \in S$,

$$\xi < s.$$

Consider $\eta = \frac{1}{2}(s + \xi) = \frac{s+\xi}{2}$.

Now $s > \xi$ implies that

$s > \frac{s+\xi}{2}$. Thus $s > \xi$.

Also $\frac{s+\xi}{2}$; i.e. $n > \xi$, therefore, $s > \eta > \xi$.

Therefore, it follows by Dedekind's axiom ($*$ above), that $\eta \in R$. Therefore by definition of R, $\eta \geq s$, for all $s \in S$; which is clearly a contradiction, because $s > \eta$. Hence our supposition that $\xi \notin R$ must be wrong. Therefore $\xi \in R$ and by the definition of R it follows that

$$\xi \geq s, \text{ for all } s \in S.$$

Since ξ is a dividing number, therefore, for an arbitrary small positive number ϵ, $\xi - \epsilon < \xi$, the dividing number. Therefore, by Dedekind's axiom ($*$ above) it follows that

$$\xi - \epsilon < s \text{ for at least one } s \in S.$$

Thus from above we have proved that there exists a number ξ such that

(1) $s \leq \xi$, for all $s \in S$.

(2) $\xi - \epsilon < s$, for at least one $s \in S$.

Therefore it follows that ξ is the least upper bound of S. Hence the result. $\qquad\square$

Archimedean Property of real numbers

Theorem 1.12. *If $a > 0$, then for any $b \in \mathbb{R}$, there exists $n \in \mathbb{N}$ such that $na > b$.*

Proof. If $b \leq 0$, then there is nothing to prove. For $b > 0$, if possible suppose $na \leq b$, for all $n \in \mathbb{N}$. Let $E = \{na \mid n \in \mathbb{N}\}$, which is nonempty and bounded above; because $na \leq b$. Therefore the least upper bound of E exists. Let l be the least upper bound of E. Therefore, $l - a$ can not be an upper bound of E. So, there exists a positive integer n such that $na > l - a$; i.e. $(n + 1)a > l$. But $(n + 1)a \in E$. Therefore, a is not a least upper bound of E, which is a contradiction. Hence our supposition must be wrong. Therefore there must exist a positive integer n such that $na > b$. $\qquad\square$

Corollary 1.13. *For any $a \in \mathbb{R}$, there exists $n \in \mathbb{N}$ such that $n > a$.*

Proof. The proof is obvious from Theorem (1.12) $\qquad\square$

Corollary 1.14. *For any $a \in \mathbb{R}$, there exists an integer m such that $m < a$.*

Proof. Corollary (1.13) implies that there exists a positive integer n such that $n > -a$. Taking $m = -n$, it follows that $m < a$. $\qquad\square$

Corollary 1.15. *For any positive real number a there exists $n \in \mathbb{N}$ such that $a > \frac{1}{n}$.*

Proof. Suppose a is a positive real number then $\frac{1}{a}$ is also a positive real number. Therefore by Corollary (1.13) it follows that there exists a positive integer n such that $n > \frac{1}{a}$. Hence $a > \frac{1}{n}$. $\qquad\square$

Lemma 1.16. *If x is any real number, then there exists an integer m such that $x - 1 \leq m < x$.*

Proof. Let x be any real number. The by Corollary (1.13) it follows that there exists an integer less than x. Let m be the such that $m < x \leq m+1$. This implies $x - 1 \leq m < x$. $\qquad\square$

Theorem 1.17. (Rational Density Theorem) *Between any two distinct real numbers there always exists a rational number.*

Proof. Let a and b be any two real numbers such that $a < b$; *i.e.* $b - a > 0$. So, by Corollary (1.15) it follows that there exists a positive integer n such that $\frac{1}{n} < b - a$ or

$$a < b - \frac{1}{n} \tag{4}$$

Consider the real number $x = nb$ Then by above lemma $x - 1 \leq m < x$, for some integer m. This implies $nb - 1 \leq m < nb$, or $b - \frac{1}{n} \leq \frac{m}{n} < b$. Therefore, by (4) $a < b - \frac{1}{n} \leq \frac{m}{n} < b$; i.e. $a < \frac{m}{n} < b$. Since n is a positive integer, $\frac{m}{n}$ is a well defined rational number. Thus there exists a rational number between two distinct real numbers. $\qquad\square$

Corollary 1.18. *Between any two distinct real numbers there exist an infinite number of rational numbers.*

Proof. Let a and b be any two numbers such that $a < b$. Now, by the above theorem there lies a rational number r_1 between a and b. Now r_1 being a rational is real and a is also a real number. so, by the same theorem it follows that there lies a rational number r_2 between a and r_1. By the repeated application of this theorem it follows that there exist an infinite number of rational numbers between a and b. $\qquad\square$

Theorem 1.19. (Irrational density theorem) *Between any two distinct real numbers there always exists an irrational number.*

Proof. Let a and b be any two real numbers such that $a < b$. So, $a - \sqrt{2} \leq b - \sqrt{2}$. So, by rational density theorem there exists a rational number

r (say) such that $a - \sqrt{2} < r \leq b - \sqrt{2}$; i.e. $a < r + \sqrt{2} < b$. Now $r + \sqrt{2}$ being an irrational implies that there exists an irrational number between a and b. Hence the proof. $\qquad\square$

Corollary 1.20. *Between any two distinct real numbers there exist an infinite number of irrational numbers.*

Proof. Let a and b be the any two real numbers such that $a < b$. Then by irrational density theorem (1.19), there exists an irrational number r_1 (say) between a and b. Now r_1 being an irrational is real and a is also a real number, therefore, by the same theorem it follows that there exists an irrational number r_2 (say) between a and r_1. By the repeated application of this theorem it follows that there exist an infinite number of irrational numbers between a and b. $\qquad\square$

Remark 1.21. Suppose a and b are distinct real numbers say $a < b$. In order to find a real number between a and b we make use of obvious inequality $a < \frac{(a+b)}{2} < b$. Thus $\frac{(a+b)}{2}$ is a real number lying between a and b. In this way we can find an infinite number of real numbers between any two real numbers. But, if we are asked to prove the existence of a rational or an irrational number between any two distinct real numbers, then this proof is not sufficient.

Remark 1.22. Irrational numbers can be approximated by rational numbers as close as we like. For example $\sqrt{2}$ is an irrational number and $1 < \sqrt{2}$. 1 and $\sqrt{2}$ are real numbers, so by rational density theorem (8.4), there exists a rational number r_1 (say) between 1 and $\sqrt{2}$. Now r_1 and $\sqrt{2}$ are real numbers. Therefore, by the same theorem it follows that there lies a rational number r_2 (say) between r_1 and $\sqrt{2}$. S0, $1 < r_1 < r_2 < \sqrt{2}$. By the repeated application of this theorem, it follows that the irrational number $\sqrt{2}$ can be approximated very closely by the rational numbers.

Definition 1.23. Absolute value of a real number x is denoted by $\mid x \mid$ and is defined as

$$\mid x \mid = x \text{ if } x \geq 0, \text{ and}$$

$$\mid x \mid = -x \text{ if } x < 0.$$

For example, $\left|\frac{-3}{4}\right| = \frac{3}{4}$ and $|\,5\,| = 5$.

Theorem 1.24. (Triangle inequality) *If x and y are real numbers, then*

(1) $|\,x+y\,| \leq |\,x\,| + |\,y\,|$

(2) $|\,x-y\,| \geq |\,|\,x\,| - |\,y\,|\,|$

Proof. **(1)** We have
$|\,x+y\,| = x+y$ if $x+y \geq 0$ and
$|\,x+y\,| = -(x+y)$ if $x+y < 0$.
But $x \leq |\,x\,|$ for all x and $y \leq |\,y\,|$ for all y. So, $|\,x+y\,| = x+y$ if $x+y \geq 0$ implies that $|\,x+y\,| \leq |\,x\,| + |\,y\,|$ for all $x+y \geq 0$. Also $|\,x+y\,| = -(x+y)$ if $x+y < 0$, which implies that $|\,x+y\,| = -x-y$ if $x+y < 0$. But $-x \leq |\,x\,|$ for all x and $-y \leq |\,y\,|$ for all y. So, $|\,x+y\,| \leq |\,x\,| + |\,y\,|$. Therefore it follows that if x and y are any real numbers, then $|\,x+y\,| \leq |\,x\,| + |\,y\,|$.

(2) We have $|\,x\,| = |\,(x-y)+y\,|$. So, by part (1) above $|\,x\,| \leq |\,x-y\,| + |\,y\,|$, which implies that

$$|\,x\,| - |\,y\,| \leq |\,x-y\,|. \tag{5}$$

Similarly $|\,y\,| - |\,x\,| \leq |\,y-x\,|$. But $|\,x-y\,| = |\,y-x\,|$. Therefore

$$|\,y\,| - |\,x\,| \leq |\,x-y\,|. \tag{6}$$

From (5) and (6), we get $|\,x-y\,| \geq |\,|\,x\,| - |\,y\,|\,|$. $\qquad\square$

EXERCISES

(1) If x and y are integers and $x - y$ is even. Show that $x^2 - y^2$ is divisible by 4.

(2) If x is an even integer, then show that x^2 is also an even integer.

(3) Define rational number. If x and y are rational numbers, then prove that $x + y$, $x - y$, xy and $\frac{x}{y}$ where $y \neq 0$ are rational numbers.

(4) Let $x\sqrt{2} + y\sqrt{3} = 0$ where x and y are both rational numbers. Show that $x = 0 = y$.

(5) Show that following numbers are not rational

    (a) $\sqrt{2}$

    (b) $2 + \sqrt{3}$

    (c) $\sqrt{3}\sqrt{2}$

    (d) $\log 3$

(6) state the law of trichotomy for real numbers.

(7) Represent geometrically the real numbers

    (a) $\frac{3}{5}$

    (b) $\frac{-7}{4}$

    (c) $\sqrt{3}$

(8) Define magnitude (modulus) of a real number. Prove that for two real numbers x and y,

    (a) $|x + y| \leq |x| + |y|$

    (b) $||x|| - |y|| \leq |x + y|$

(9) Prove that for real numbers $x_1, x_2, ..., x_n$

    (a) $|x_1 + x_2 + ... + x_n| \leq |x_1| + |x_2| + ... + |x_n|$

    (b) $|x_1 x_2 ... x_n| = |x_1||x_2|...|x_n|$

(10) Use Mathematical induction to prove that

(a) $\frac{1}{2} \cdot \frac{3}{4} \cdots \frac{2n-1}{n} < \frac{1}{\sqrt{2n+1}}$

(b) $n^{n+1} > (n+1)^n$

(c) $(1 + 2 + \ldots + n)^2 = 1^3 + 2^2 + \ldots + n^3$

(11) If x and y are real numbers, show that

$$\frac{|x+y|}{1+|x+y|} \le \frac{|x|}{1+|x|} + \frac{|y|}{1+|y|}$$

(12) If x and y are real numbers, then

(a) $|x+y| = |x| + |y|$ if and only if $xy \ge 0$

(b) $|x+y| < |x| + |y|$ if and only if a $xy < 0$

(13) Show that union of two bounded sets is bounded.

(14) Give three examples of sets which are

(a) bounded

(b) not bounded

(15) Which of the following sets are bounded

(a) $\{1, 3, 5, 7, 9\}$

(b) $\{\frac{1}{n}; n \in \mathbb{N}\}$

(c) $\{\frac{(-1)^n}{n}; n \in \mathbb{N}\}$

(16) Show that greatest (or the smallest) member of a set (in case it exists) is unique.

(17) Give an example of a set having supremum equal to infimum.

(18) If $S \subseteq T \subseteq R$, where $S \ne \phi$, then show that

(a) T is bounded above implies that $sup(S) \le sup(T)$

(b) T is bounded below implies that $inf(T) \le inf(S)$

(19) Find the infimum and the supremum of the following sets. Which of these belong to the set.

(a) $\{-1, \frac{-1}{2}, \frac{-1}{3}, \ldots\}$

    (b) $\{-2, \frac{-3}{2}, \frac{-4}{3}, ..., -\frac{n+1}{n}\}$

    (c) $]a, b[$

    (d) $\{1 + \frac{(-1)^n}{n} : n \in \mathbb{N}\}$

(20) Which of the sets in above question are bounded.

(21) Find $sup(S)$ and $inf(S)$, where

    (a) $S = \{x \in R;\ x^2 > 2\}$

    (b) $S = (0, 1)$

    (c) $S = \{\frac{m}{n};\ m, n \in \mathbb{Z},\ n \neq 0\}$ iv) $S = \{\frac{1}{m} + \frac{1}{n};\ m, n \in \mathbb{N}\}$

    (d) $S = \{1, 4, 9, 16, ...\}$

Chapter2

SEQUENCES AND THEIR CONVERGENCE

George Cantor (1845 - 1918), the creator of set theory, made considerable contributions to the development of theory of real sequences. The study of many important and advanced concepts become easy if the notion of sequences is employed.

Sequences and their convergence form an important component of mathematical analysis and arise in many situations. It has been provided by George cantor and A. Cauchy that sequences and their convergence are essentially the most effective tools of mathematics.

Definition 2.1. A (real) sequence $\{a_n\}_1^\infty$, where $n \in \mathbb{N}$ is defined as a mapping from the set of positive integers into the set of real numbers. The set of positive integers is called the domain of the sequence and the set of real numbers $\{a_1, a_2, a_3, ...\}$ is called the range of the sequence.

For example; consider the sequence $\{a_n\}_1^\infty$, where $a_n = n, \frac{1}{n}, 1 + \frac{1}{n}, (-1)^n$. The ranges of first, second, third and the fourth sequences are $\{1, 2, 3, ..., n, ...\}$; $\{1, \frac{1}{2}, \frac{1}{3}, ..., \frac{1}{n}, ...\}$; $\{2, \frac{3}{2}, \frac{4}{3}, ..., 1 + \frac{1}{n}, ...\}$ and $\{-1, 1\}$ respectively. From these examples it follows that a sequence can be finite or infinite; bounded or unbounded. The sequence $\{-1, 1\}$ is finite and bounded. The sequences $\{1, \frac{1}{2}, \frac{1}{3}, ..., \frac{1}{n}, ...\}$; $\{2, \frac{3}{2}, \frac{4}{3}, ..., 1 + \frac{1}{n}, ...\}$ are infinite but bounded. The sequence $\{1, 2, 3, ..., n, ...\}$ is infinite and unbounded.

Definition 2.2. Convergence of a sequence: A sequence $\{a_n\}_1^\infty$ is said to converge to a or is to have limit a if given any $\epsilon > 0$, how so ever small, there exists a positive integer N such that $\mid a_n - a \mid \leq \epsilon$ for all $n \geq N$ and we write

$$lim_{n \longrightarrow \infty} a_n = a.$$

Thus a sequence $\{a_n\}_1^\infty$ converges to a if the distance between a_n and a becomes smaller and smaller as we increase n.

Example 2.3. Consider the sequence $\{a_n\}_1^\infty$, $a_n = \frac{1}{n}$, where $n \in \mathbb{N}$. It is the sequence $\{1, \frac{1}{2}, \frac{1}{3}, ..., \frac{1}{n}, ...\}$. Let us choose $\epsilon > \frac{3}{100}$. Then $\mid a_n - 0 \mid <$ $\frac{3}{100}$ or $\mid \frac{1}{n} - 0 \mid < \frac{3}{100}$, which is true if $\frac{1}{n} < \frac{3}{100}$; i.e. if $n > \frac{100}{3} = 33 + (\frac{1}{3})$; i.e. if $n \geq 34$. Thus in this case $N = 34$.

Now suppose that $\epsilon = \frac{3}{1000}$. Then $\mid a_n - 0 \mid < \frac{3}{1000}$

or, $\mid \frac{1}{n} - 0 \mid < \frac{3}{1000}$,

which is true if $\frac{1}{n} \frac{3}{1000}$; i.e. if $n > \frac{1000}{3} = 333 + (\frac{1}{3})$; i.e. if $n \geq 334$. In this case $N = 334$. Thus if $\epsilon = \frac{3}{100}$, then $N = 34$ and if $\epsilon = \frac{3}{1000}$, then $N = 334$.

Remark 2.4. Every convergence sequence is bounded.

Proof. Suppose a sequence $\{a_n\}_1^\infty$ converges to a. Then given any $\epsilon > 0$, how so ever small, there exists a positive integer N such that $\mid a_n - a \mid \leq \epsilon$, for all $n \geq N$, which implies that

$$a_n - a \leq \epsilon \tag{7}$$

and

$$-(a_n - a) \leq \epsilon, i.e. a_n - a \geq -\epsilon \tag{8}$$

From (7) and (8), we get

$$-\epsilon \leq a_n - a \leq \epsilon.$$

This shows the sequence $\{a_n\}_1^\infty$ is bounded. $\qquad\qquad\square$

Theorem 2.5. *Prove that the limit of a convergent sequence is unique.*

Proof. Consider a sequence $\{a_n\}_1^\infty$. Suppose it converges to a; i.e. $lim_{n\longrightarrow\infty}a_n = a$. Therefore, given any $\epsilon > 0$, how so ever small, there exists a positive integer N such that for all $n \geq N$

$$| a_n - a | \leq \epsilon \tag{9}$$

We have to show that the limit of this sequence is unique. If possible suppose that the limit is not unique. Let $lim_{n\longrightarrow\infty}a_n = b$, $a \neq b$. Therefore, given any $\epsilon > 0$ how so ever small, there exists a positive M such that for all $n \geq M$

$$| a_n - a | \leq \epsilon \tag{10}$$

Since $a \neq b$, $| a - b | > 0$. We choose $\epsilon = | \frac{(a-b)}{2} |$.

Then from (9) and (10), we get for all $n \geq N$

$$| a_n - a | \leq | \frac{(a - b)}{2} | \tag{11}$$

and for all $n \geq M$

$$| a_n - b | \leq | \frac{(a - b)}{2} | . \tag{12}$$

Let L be $Max(N, M)$. Then from (11) and (12), we get for all $n \geq L$

$$| a_n - a | \leq | \frac{(a - b)}{2} | \tag{13}$$

and for all $n \geq L$

$$| a_n - b | \leq | (a - b)/2 | \tag{14}$$

Now for all $n \geq L$, $| a - b |$

$$= | (a - a_n) + (a_n - b) |$$

$$\leq |\, a_n - a \,| + |\, a_n - b \,|$$

$$< |\,|\, \tfrac{(a-b)}{2} \,| + |\, \tfrac{(a-b)}{2} \,|\,|$$

$$= |\, a - b \,|; \text{ using (13) and (14)}$$

Therefore, $|\, a - b \,| < |\, a - b \,|$, which is a contradiction. Hence our supposition must be wrong, and therefore, it follows that the limit of a convergent sequence is unique. $\qquad\qquad\square$

Theorem 2.6. *Suppose $\{a_n\}_1^\infty$ and $\{b_n\}_1^\infty$ are two convergent sequences. Then*

(1) $\lim_{n\longrightarrow\infty}(a_n + b_n) = \lim_{n\longrightarrow\infty} a_n + \lim_{n\longrightarrow\infty} b_n$

(2) $\lim_{n\longrightarrow\infty}(a_n - b_n) = \lim_{n\longrightarrow\infty} a_n - \lim_{n\longrightarrow\infty} b_n$

(3) $\lim_{n\longrightarrow\infty}(a_n . b_n) = \lim_{n\longrightarrow\infty} a_n \, . \, \lim_{n\longrightarrow\infty} b_n$

(4) $\lim_{n\longrightarrow\infty}(\tfrac{a_n}{b_n}) = \dfrac{\lim_{n\longrightarrow\infty} a_n}{\lim_{n\longrightarrow\infty} b_n} \text{ where } \lim_{n\longrightarrow\infty} b_n \neq 0.$

Proof. It is given that $\{a_n\}_1^\infty$ and $\{b_n\}_1^\infty$ are two convergent sequences, let $a_n \to a$ and $b_n \to b$. So, given any $\epsilon > 0$, how so ever small, there exists a positive integer N such that for all $n \geq N$

$$|\, a_n - a \,| \leq \epsilon \qquad\qquad\qquad (15)$$

and given any $\epsilon > 0$, how so ever small, there exists a positive M such that for all $n \geq M$

$$|\, b_n - b \,| \leq \epsilon \qquad\qquad\qquad (16)$$

Let L be $Max(N, M)$. Then from (15) and (16), we get that

$$|\, a_n - a \,| < \epsilon \qquad\qquad\qquad (17)$$

for all $n \geq L$ and

$$|\, b_n - b \,| < \epsilon \qquad\qquad\qquad (18)$$

we have the following:

(1) Now for all $n \geq L$

$$| (a_n + b_n) - (a + b) |$$
$$= | (a_n - a) + (b_n - b) |$$

$$\leq | (a_n - a) | + | (b_n - b) | \text{ (using triangle inequality)}.$$

Now using (17) and (18), we get

$$| (a_n + b_n) - (a + b) | < \epsilon + \epsilon = 2\epsilon.$$

Therefore, $| (a_n + b_n) - (a + b) | < 2\epsilon$ for all $n \geq L$. Since ϵ is arbitrary, so is 2ϵ. Hence it follows that $lim_{n \longrightarrow \infty}(a_n + b_n) = a + b = lim_{n \longrightarrow \infty}a_n + lim_{n \longrightarrow \infty}b_n$.

(2) $| (a_n - b_n) - (a - b) |$

$$= | (a_n - a) + (b - b_n) |$$

$$\leq | (a_n - a) | + | (b - b_n) | \text{ (using triangle inequality)}$$

$$\text{or, } | (a_n - b_n) - (a - b) |$$

$$\leq | (a_n - a) | + | (b_n - b) |.$$

Now using (17) and (18), we get

$$| (a_n + b_n) - (a + b) | < \epsilon + \epsilon = 2\epsilon.$$

Therefore $| (a_n - b_n) - (a - b) | < 2\epsilon$ for all $n \geq L$. Since ϵ is arbitrary so is 2ϵ. Hence it follows that $lim_{n \longrightarrow \infty}(a_n - b_n) = a - b = lim_{n \longrightarrow \infty}a_n - lim_{n \longrightarrow \infty}b_n$.

(3) $| (a_n.b_n) - (a.b) |$

$$= | a_n(b_n - b) + b(a_n - a) |$$

$$\leq | a_n(b_n - b) | + | b(a_n - a) | \text{ (using triangle inequality)}$$

$$\text{or, } | (a_n.b_n) - (a.b) |$$

$$\leq | a_n || (b_n - b) | + | b || (a_n - a) |$$

Now using (17) and (18), we get

$$\mid (a_n.b_n) - (a.b) \mid = \mid a_n \mid \epsilon + \mid b \mid \epsilon \text{ for all } n \geq L.$$

Since $\{a_n\}_1^\infty$ is a convergent sequence, therefore, it is bounded. Let $\mid a_n \mid \leq k$, where k is a constant. Therefore $\mid (a_n.b_n) - (a.b) \mid = (k + \mid b \mid)\epsilon = l\epsilon$ for all $n \geq L$, where $l = (k + \mid b \mid)$ is a constant. Since ϵ is arbitrary, so is $l\epsilon$. Hence it follows that $lim_{n\longrightarrow\infty}(a_n.b_n) = a.b = lim_{n\longrightarrow\infty}a_n.lim_{n\longrightarrow\infty}b_n$.

(4) Given $lim_{n\longrightarrow\infty}b_n \neq 0$; i.e. $b \neq 0$. Since $b_n \to b$ implies $\frac{1}{b_n} \to \frac{1}{b}$. So, using (3) above, we get

$$lim_{n\longrightarrow\infty}\left(\frac{a_n}{b_n}\right)$$

$$= lim_{n\longrightarrow\infty}\left[a_n.\left(\frac{1}{/b_n}\right)\right]$$

$$= lim_{n\longrightarrow\infty}a_n. \; lim_{n\longrightarrow\infty}\frac{1}{b_n}$$

$$= a.\left(\frac{1}{b}\right) = \frac{a}{b}$$

$$= \frac{lim_{n\longrightarrow\infty}a_n}{lim_{n\longrightarrow\infty}b_n}.$$

Hence $lim_{n\longrightarrow\infty}\frac{(a_n}{b_n}) = \frac{lim_{n\longrightarrow\infty}a_n}{lim_{n\longrightarrow\infty}b_n}$, where $lim_{n\longrightarrow\infty}b_n \neq 0$. $\square$

Exercise 2.7. If the sequence $\{a_n\}_1^\infty$ is given by $a_n = (-1)^n$, n = 1, 2, 3, ..., n, ... Prove that $lim_{n\longrightarrow\infty}a_n$ does not exist; i.e. the sequence $\{a_n\}_1^\infty$ is not convergent.

Solution: The sequence $\{a_n\}_1^\infty$ is given by $a_n = (-1)^n$, $n = 1, 2, 3, ..., n, ...$ We have to prove that $lim_{n\longrightarrow\infty}a_n$ does not exist. If possible suppose $lim_{n\longrightarrow\infty}a_n$ exists. Let $lim_{n\longrightarrow\infty}a_n = a$. So, given any $\epsilon > 0$, how so ever small, there exists a positive integer N such that for all $n \geq N$

$$\mid a_n - a \mid < \epsilon$$

or for all $n \geq N$.

$$| (-1)^n - a | < \epsilon \tag{19}$$

If n is odd, then from (19), $| -1 - a | < \epsilon$, for all $n \geq N$

or, $| -1(1 + a) | < \epsilon$

or, $| (1 + a) | < \epsilon$, for all $n \geq N$.

If n is even, then from (19) we get $| 1 - a | < \epsilon$ for all $n \geq N$. Now $| (1 + a) | < \epsilon$, and $| (1 - a) | < \epsilon$, for all $n \geq N$ and for any ϵ, taking in particular $\epsilon = 1$, we get $| (1 + a) | < 1$ and $| (1 - a) | < 1$.

Now $2 = | 1 + 1 | = | (1 + a) + (1 - a) | \leq | (1 + a) | + | (1 - a) | < 1 + 1 = 2$, which implies that $2 < 2$, a contradiction. Hence our supposition that $lim_{n \longrightarrow \infty} a_n$ exist is wrong, and we conclude that $lim_{n \longrightarrow \infty} a_n$ does not exist.

Remark 2.8. In the above example $a_n = \{-1, 1\}$ is bounded but not convergent. So, a bounded sequence need not to be convergent.

Definition 2.9. For any $a, b \in \mathbb{R}$ with $a \neq b$ open interval $(a, b) = \{x \mid a < x < b\}$, and closed interval $[a, b] = \{x \mid a \leq x \leq b\}$.

Definition 2.10. Let h be a positive number. Then the neighborhood of x is the open interval $(x - h, x + h)$.

Definition 2.11. Let E be an infinite set. An element x (may or may not in E) is said to be the limit point of E if every neighborhood of x contains an infinite number of elements of E.

For example let $E = \{1, \frac{1}{2}, \frac{1}{3}, ...\}$. The semi closed interval $[0, \frac{1}{100})$ is a neighborhood of 0 and contains an infinite number of elements $\frac{1}{100}, \frac{1}{101}, \frac{1}{102}, ...$ of E.

Theorem 2.12. *If E is an infinite set and x is a limit point of E, then there exists a sequence in E which converges to x.*

Proof. Let E be an infinite set and x a limit point of E. Then, every neighborhood of x contains at least one point of E. Let us take a neighborhood of unit length. Then there exists an element $x_1 \in E$ such that

$$| x - x_1 | < 1.$$

Similarly there exists $x_2 \in E$, $x_1 \neq x_2$ such that

$$| x - x_2 | < \tfrac{1}{2}$$

Again there exists $x_3 \in E$, $x_2 \neq x_3$ such that

$$| x - x_3 | < \tfrac{1}{3}$$

Repeating the process n-times, we get that there exists $x_n \in E$, $x_{n-1} \neq x_n$ such that
$$| x - x_n | < \frac{1}{n} \tag{20}$$

Repeating the process indefinitely, we obtain a sequence $\{x_n\}_1^\infty$, where $x_n \in E$ for all $n = 1, 2, 3, \ldots$. Let ϵ be any positive arbitrary number.

Then $\frac{1}{n} < \epsilon$ if $1 < n\epsilon$, i.e. $n\epsilon > 1$ or $n > \frac{1}{\epsilon}$.

Choose $N = \frac{1}{\epsilon} + 1$. Now if $n \geq N$, then $n > \frac{1}{\epsilon}$. So, $\frac{1}{n} < \epsilon$. Hence from (20) it follows that $| x - x_n | < \epsilon$, for all $n \geq N \Rightarrow \lim_{n \to \infty} x_n = x$. This proves the result. $\square$

Definition 2.13. A sequence $\{a_n\}_1^\infty$ is called monotonic increasing if

$$a_1 \leq a_2 \leq a_3 \leq \ldots \leq a_n \leq a_{(n+1)} \leq \ldots$$

Definition 2.14. A sequence $\{a_n\}_1^\infty$ is called monotonic decreasing if

$$a_1 \geq a_2 \geq a_3 \geq \ldots \geq a_n \geq a_{(n+1)} \geq \ldots$$

Theorem 2.15. *Every monotonic increasing sequence which is bounded above is convergent.*

Proof. Let $\{a_n\}_1^\infty$ be a monotonic increasing sequence which is bounded above. Since the sequence is monotonic increasing, we have

$$a_1 \leq a_2 \leq a_3 \leq \dots \leq a_n \leq a_{(n+1)} \leq \dots$$

Now E is non empty and bounded above, therefore, the least upper bound of E exists. Let a be the least upper bound of E. Therefore, $a_n \leq a$ and

$$a_n < a + \epsilon \tag{21}$$

where ϵ is an arbitrary positive number.

Again as a is the least upper bound of E, therefore, given $\epsilon > 0$, there must exist a positive integer N such that

$$a_N > a - \epsilon \tag{22}$$

If $n \geq N$ then

$$a_n \geq a_N \tag{23}$$

Now the sequence $\{a_n\}_1^\infty$ is monotonic increasing, therefore, from (22) and (23), we get

$a_n > a - \epsilon.$

Therefore from (21) and (23), we get

$a - \epsilon < a_n < a + \epsilon$, for all $n \geq N$

or, $a - \epsilon - a < a_n - a < a + \epsilon - a$, for all $n \geq N$

or, $-\epsilon < a_n - a < \epsilon$, for all $n \geq N$

or, $\mid a_n - a \mid < \epsilon$, for all $n \geq N$,

which implies that $\lim_{n \to \infty} a_n = a$. Hence it follows that the sequence $\{a_n\}_1^\infty$ is convergent. $\square$

Theorem 2.16. *Every monotonic decreasing sequence which is bounded below is convergent.*

Proof. Let $\{a_n\}_1^\infty$ be a monotonic decreasing sequence which is bounded below. Now the sequence is monotonic decreasing, therefore,

$$a_1 \geq a_2 \geq a_3 \geq \ldots \geq a_n \geq a_{(n+1)} \geq \ldots$$

Also E is non empty and bounded below, therefore, the greatest lower bound of E exists. Let a be the greatest lower bound of E. Therefore, $a_n \geq a$. So,

$$a_n > a - \epsilon \tag{24}$$

where ϵ is an arbitrary positive number. Again as a is the least upper bound of E, therefore, given $\epsilon > 0$, there must exist an integer N such that

$$a_N < a + \epsilon \tag{25}$$

If $n \geq N$, then

$$a_N \geq a_n. \tag{26}$$

Since the sequence $\{a_n\}_1^\infty$ is monotonic decreasing, from (24) and (25), we get

a $-\epsilon < a_n < a + \epsilon$ for all $n \geq N$

or, a $-\epsilon - a < a_n - a < a + \epsilon - a$, for all $n \geq N$

or, $-\epsilon < a_n - a < \epsilon$, for all $n \geq N$

or, $\mid a_n - a \mid < \epsilon$, for all $n \geq N$,

which implies that $lim_{n \to \infty} a_n = a$. Hence it follows that the sequence $\{a_n\}_1^\infty$ is convergent. $\square$

Exercise 2.17. Prove that $lim_{n \to \infty}(1 + \frac{1}{n})^n$ exists and lies between 2 and 3.

Solution: Suppose $a_n = \lim_{n \to \infty}(1 + \frac{1}{n})^n$, $n = 1, 2, 3, \ldots$. Consider $(n+1)$ numbers; $1, (1 + \frac{1}{n}), (1 + \frac{1}{n}), (1 + \frac{1}{n}), \ldots, (1 + \frac{1}{n})$ (n-times). Now the arithmetic mean is always greater than geometric mean, therefore, it follows that

$$\frac{1 + (\frac{1}{n}) + (\frac{1}{n}) + \ldots + (\frac{1}{n})}{n+1} > [1.(1 + \frac{1}{n}).(1 + \frac{1}{n})\ldots(1 + \frac{1}{n})]^{\frac{1}{1+n}}.$$

This implies that

$$\frac{1 + n(1 + \frac{1}{n})}{(n+1)} > [(1 + \frac{1}{n})^n]^{\frac{1}{1+n}}$$

or, $\frac{(1 + n + 1)}{n+1} > (1 + \frac{1}{n})^{\frac{n}{1+n}}$

i.e. $\frac{(1)}{n+1} > (1 + \frac{1}{n})^{\frac{n}{1+n}}$

i.e. $[1 + \frac{1}{n+1}]^{n+1} > (1 + \frac{1}{n})^n$

which implies that $a_{n+1} > a_n$ for all $n = 1, 2, 3, \ldots$.

Hence it follows that the sequence $\{a_n\}_1^\infty$ is monotonic increasing.

Now $(1 + \frac{1}{n})^n$

$$= 1 + n.\frac{1}{n} + [\frac{n(n-1)}{2!}](\frac{1}{n})^2 + [\frac{n(n-1)(n-2)}{3!}](\frac{1}{n})^3 + \ldots$$
$$+ \frac{[n(n-1)(n-2)\ldots(n-n+1)}{n!}](\frac{1}{n})^n,$$

which implies that

$$(1 + \frac{1}{n})^n = 2 + (1 - \frac{1}{n})(\frac{1}{2!}) + (1 - \frac{1}{n})(1 - \frac{2}{n})(\frac{1}{3!}) + \ldots$$
$$+ [(1 - \frac{1}{n})(1 - \frac{2}{n})]\ldots[\frac{1 - (n-1)}{n}](\frac{1}{n!})$$

i.e. $(1 + \frac{1}{n})^n > 2$ for all $n = 1, 2, 3, \ldots$.

Therefore, for all $n = 1, 2, 3, \ldots$.

$$a_n > 2 \tag{27}$$

But $(1 - \frac{1}{n}) < 1$ so, we have;

$(1 + \frac{1}{n})^n$

$$< 2 + \tfrac{1}{2!} + \tfrac{1}{3!} + \dots + \tfrac{1}{n!}$$

$$< 2 + \left(\tfrac{1}{2} + \tfrac{1}{4} + \tfrac{1}{8} + \dots\right)$$

$$< 2 + \frac{\left(\tfrac{1}{2}\right)}{\left(1 - \tfrac{1}{2}\right)} = 3.$$

So, $(1 + \tfrac{1}{n})^n < 3$, for all $n = 1, 2, 3, \dots$;

i.e. for all $n = 1, 2, 3, \dots,$

$$a_n < 3 \tag{28}$$

Hence it follows that the sequence $\{a_n\}_1^\infty$ is bounded above. Thus the sequence $\{a_n\}_1^\infty$ monotonic increasing and bounded above. Therefore it follows that $\{a_n\}_1^\infty$ is convergent. Hence $lim_{n \to \infty} a_n$ exists, and from (27) and (28) we have
$2 < lim_{n \to \infty}(1 + \tfrac{1}{n})^n < 3$. Hence proved.

Note: It is customary to denote $lim_{n \to \infty}(1 + \tfrac{1}{n})^n$ by e. The above example shows that $2 < e < 3$. The usual definition of e is given by

$$e = \sum_{n=0}^{\infty} \tfrac{1}{n!}$$

and is an irrational number.

Theorem 2.18. (Nested Interval Theorem) *Suppose $I_n = [a_n, b_n]$, $n = 1, 2, 3, \dots$ such that*
$I_1 \supseteq I_2 \supseteq \dots \supseteq I_n \supseteq I_{n+1} \supseteq \dots$
and length of I_n; i.e. $L(I_n) \to 0$ as $n \to \infty$, then there exists an element x such that $x \in I_n$, for all $n = 1, 2, 3, \dots$.

Proof. Let $I_n = [a_n, b_n]$, $n = 1, 2, 3, \dots$ such that $I_1 \supseteq I_2 \supseteq \dots \supseteq I_n \supseteq I_{n+1} \supseteq \dots$ and length of I_n; i.e. $L(I_n) \to 0$ as $n \to \infty$. Since $I_1 \supseteq I_2 \supseteq I_3 \supseteq \dots$, it follows that $a_1 \leq a_2 \leq a_3 \leq \dots$; i.e. the sequence $\{a_n\}_1^\infty$ is monotonic increasing and also bounded above by

b_1. Thus the sequence $\{a_n\}_1^\infty$ is a monotonic increasing and bounded above. Therefore, it follows that the sequence $\{a_n\}_1^\infty$ is convergent, say $lim_{n\to\infty} a_n = a$.

Now $I_1 \supseteq I_2 \supseteq I_3 \supseteq ...$, therefore, it follows that $b_1 \geq b_2 \geq b_3 \geq ...$; i.e the sequence $\{b_n\}_1^\infty$ is monotonic decreasing and also bounded below by a_1. Thus the sequence $\{b_n\}_1^\infty$ is monotonic decreasing and bounded below. Therefore, it follows that the sequence $\{b_n\}_1^\infty$ is convergent, say $lim_{n\to\infty} b_n = b$. Given that $L(I_n) \to 0$ as $n \to \infty$. But $L(I_n) = b_n - a_n$; i.e. $lim_{n\to\infty} b_n - a_n = 0$. This implies that $lim_{n\to\infty} b_n - lim_{n\to\infty} a_n = 0$; i.e. $b - a = 0$ or $b = a$. Let $b = a = x$. Now $a_n \leq a$ and $b_n \leq b$, therefore, $a_n \leq x$ and $b_n \leq x$; i.e. $a_n \leq x \leq b_n$. So, $x \in [a_n, b_n]$, $n = 1, 2, 3, ...$ Thus $x \in I_n$ for all n. $\qquad\square$

Theorem 2.19. *Every infinite bounded set has at least one limit point.*

Proof. Let E be an infinite bounded set. Since E is bounded so, there exist real numbers a and b such that $E \subseteq [a, b] = I$, where $a < b$. Bisect $[a, b]$ and consider two subintervals $[a, \frac{(a+b)}{2}]$ and $[\frac{(a+b)}{2}, b]$. Since E is infinite set so, one of the subintervals will contain an infinite number of elements of E. Let $I_1 = [a, \frac{(a+b)}{2}]$ be that subinterval. Similarly bisect I_1 into two sub intervals and let I_2 be the subinterval of I_1 that contains an infinite number of elements of E. Therefore, $I_1 \supseteq I_2$. Again bisect I_2 into two subintervals and let I_3 be the subinterval I_2 that contain an infinite number of elements of E. Now $I_2 \supseteq I_3$. Repeating the process and at the nth stage we bisect I_{n-1} into two subintervals and let I_n be the subinterval of I_{n-1} that contains an infinite number of elements of E. Thus $I_{n-1} \supseteq I_n$. So, $I_1, I_2, I_3, ..., I_n, ...$ are closed intervals such that $I_1 \supseteq I_2 \supseteq I_3 \supseteq ... I_n \supseteq I_{n+1} \supseteq ...$ and length of I_n; i.e $L(I_n) = \frac{b-a}{2^n} \to 0$ as $n \to \infty$. Therefore $L(I_n) \to 0$ as $n \to \infty$. So, by Nested Interval Theorem there exist an element x such that $x \in I_n$ for all n. Now consider the neighborhood $(x - h, x + h)$ of x, where h is an arbitrary positive number. We choose n so large such that $I_n \subseteq (x - h, x + h)$. This is possible because $L(I_n) \to 0$ as $n \to \infty$ and $x \in I_n$ for all n. Since I_n contains an infinite number of elements of E, so from above it follows that the arbitrary neighborhood $(x - h, x + h)$ also contains an infinite number of elements of E. Hence it follows that x is a limit point of E. Thus if E is an infinite bounded set, then it has at least one limit point. $\qquad\square$

Remark 2.20. From the proof of the above theorem it follows that an infinite set can have more than one limit points.

Remark 2.21. The condition of boundedness in the above theorem is necessary; for example, if $E = \{1, 2, 3, ...\}$, then E is an infinite set but is not bounded. Now if x is a positive number then the neighborhood $(x-h, x+h)$ contains only one point namely x if x is a positive integer and it contains no point of E if x is not a positive integer. Therefore it follows that x is not a limit point of E. Thus the condition of boundedness is necessary.

Theorem 2.22. *Every bounded sequence contains a convergent subsequence.*

Proof. Suppose the sequence $\{a_n\}_1^\infty$ is bounded. Let E be the range of the sequence. Then there are two possibilities. Either E is infinite or E is finite. Now if E is infinite, then E is an infinite and bounded set. So, it has a limit point. Now E contains a limit point and E is infinite so, there will exist a sequence in E which will converge to that limit point. Thus this sequence will be convergent subsequence.

Now if E is finite, say $E = \{a_1, a_2, a_3, ..., a_k\}$, where k is a fixed positive integer, then there will be infinite number of elements in the sequence $\{a_n\}_1^\infty$ which will be equal to a_1; i.e. $ak_1 = ak_2 = ak_3 = ... = a_1$. Thus $\{a_{k_r}\}_{r=1}^\infty$ is a sequence which is convergent to a_1. Similarly there will exist another sequence which will converge to a_2 and so on. Thus if E is finite then there will exist convergent subsequences. $\qquad\square$

Exercise 2.23. Prove that $lim_{n\to\infty} n^{\frac{1}{n}} = 1$.

Solution: $(1 + \frac{1}{n})^n$

$$= 1+n.\frac{1}{n}+\left[\frac{n(n-1)}{2!}\right]\left(\frac{1}{n}\right]^2+\left[\frac{n(n-1)(n-2)}{3!}\right]\left(\frac{1}{n}\right)^3+...+\left[\frac{n(n-1)(n-2)...(n-n+1)}{1}n!\right)\left(\frac{1}{n}\right)^n$$

This implies that
$(1 + \frac{1}{n})^n$

$$= 2+(1-\tfrac{1}{n})(\tfrac{1}{2!})+(1-\tfrac{1}{n})(1-\tfrac{2}{n})(\tfrac{1}{3!})+...+\left[(1-\tfrac{1}{n})(1-\tfrac{2}{n})...1-\tfrac{(n-1)}{n}\right](\tfrac{1}{n!}).$$

But $(1 - \frac{1}{n}) < 1$ so, we have

$(1 + \frac{1}{n})^n$

$< 2 + \frac{1}{2!} + \frac{1}{3!} + ... + \frac{1}{n!}$

$< 2 + (\frac{1}{2} + \frac{1}{4} + \frac{1}{8} + ...)$

$< 2 + \frac{(\frac{1}{2})}{(1 - \frac{1}{2})} = 3.$

So, for all $n = 1, 2, 3, ...,$

$$(1 + \frac{1}{n})^n < 3, \tag{29}$$

Let $a_n = n^{\frac{1}{n}}$. Then $a_n > a_{n+1}$

if $n^{\frac{1}{n}} > (n+1)^{\frac{1}{(n+1)}}$;

i.e. if $n^{n+1} > (n+1)^n$;

i.e. if $n > \frac{(n+1)^n}{n^n}$;

i.e. if $n > (1 + \frac{1}{n})^n$, which is true by (29), if $n > 3$.

Hence it follows that $a_n > a_{n+1}$ if $n > 3$ and hence the sequence $\{a_n\}_1^\infty$ is monotonic decreasing for $n > 3$.

Now $n^{\frac{1}{n}} \geq 1$, for all $n = 1, 2, 3, ...,$

Hence it follows that $\{a_n\}_1^\infty$ is bounded below. Thus $\{a_n\}_1^\infty$ is monotonic decreasing and bounded below. Therefore, it must be convergent say

$lim_{n \to \infty} a_n = a$

i.e. $lim_{n \to \infty} n^{\frac{1}{n}} = a$

Let $x_n = n^{\frac{1}{n}} - 1$, $n = 1, 2, 3, ...$ and $x_n \geq 0$, for all $n = 1, 2, 3, ...$

or, $n^{\frac{1}{n}} = 1 + x_n$

or, $n = (1 + x_n)^n$

or, $n = 1 + nx_n + \frac{n(n-1)}{2!}x_n^2 + ... + \frac{n(n-1)(n-2)...(n(n-1))}{n!}x_n^n$

Since each term on right hand side is non negative, so

$n \geq \frac{n(n-1)}{2!}x_n^2$

or, $1 \geq \frac{(n-1)^2}{2!}x_n^2$

or, $x_n^2 \leq \frac{2}{n-1}$

or, $x_n \leq \sqrt{\frac{2}{n-1}}$

or, $0 \leq x_n \leq \sqrt{\frac{2}{n-1}}$ (because limit$(0) = 0$)

Therefore, $0 \leq lim_{n \to \infty} x_n \leq lim_{n \to \infty} \sqrt{\frac{2}{n-1}}$

or, $0 \leq lim_{n \to \infty} x_n \leq 0$

or, $lim_{n \to \infty} x_n = 0$

Therefore, $lim_{n \to \infty}((n)^{\frac{1}{n}} - 1) = 0$, which implies that

$lim_{n \to \infty} n^{\frac{1}{n}} - 1 = 0$

Hence $lim_{n \to \infty} n^{\frac{1}{n}} = 1$.

Cauchy Sequence

Definition 2.24. A sequence $\{a_n\}_1^\infty$ is said to be a Cauchy sequence if given $\epsilon > 0$, how so ever small, we can find a positive integer N such that $|a_n - a_m| < \epsilon$, for all $n, m \geq N$. In other words a sequence is a Cauchy sequence if the distance between its elements becomes very small after some stage.

Cauchy's general principle of convergence:

Theorem 2.25. *A sequence $\{a_n\}_{n=1}^{\infty}$ is convergent if and only if it is a cauchy sequence.*

Proof. Suppose $\{a_n\}_{n=1}^{\infty}$ is convergent, say $lim_{n\to\infty} a_n = a$. Therefore, given any $\epsilon > 0$ howsoever small we can find a positive integer N such that for all $n \geq N$

$$| a_n - a |< \epsilon, \tag{30}$$

and for all $m \geq N$.

$$| a_m - a |< \epsilon \tag{31}$$

Now $| a_n - a_m |$

$$=| (a_n - a) - (a_m - a) |$$

$$\leq | a_n - a | + | a_m - a | \text{ (by triangle inequality)}$$

Therefore, $| a_n - a_m |< \epsilon + \epsilon$, for all $n, m \geq N$ (by using (30) and (31))

i.e. $| a_n - a_m |< 2\epsilon$, for all $n, m \geq N$.

Hence it follows that $\{a_n\}_{n=1}^{\infty}$ is a cauchy's sequence . Thus if $\{a_n\}_{n=1}^{\infty}$ is convergent sequence, then it is a Cauchy sequence.
Conversely suppose that $\{a_n\}_{n=1}^{\infty}$ is a cauchy sequence, we will prove that it is convergent.

Since $\{a_n\}_{n=1}^{\infty}$ is a Cauchy sequence, therefore, given $\epsilon > 0$ howsoever small we can find a positive integer N such that for all $n, m \geq N$

$$| a_n - a_m |< \epsilon \tag{32}$$

We will now prove that the sequence $\{a_n\}_{n=1}^{\infty}$ is bounded.

Since (32) is true for all $n, m \geq N$, taking in particular $m = N$, we get for all $n \geq N$

$$| a_n - a_N |< \epsilon \tag{33}$$

Now $| a_n |$

$$=\mid a_n - a_N + a_N \mid$$

$$\leq \mid a_n - a_N \mid + \mid a_N \mid$$

Therefore, $\mid a_n \mid < \epsilon + \mid a_N \mid$, for all $n \geq N$ (using (33))

or, $\mid a_n \mid < M_1$, where $M_1 = \epsilon + \mid a_N \mid$

Therefore, for all $n \geq N$

$$\mid a_n \mid < M_1 \tag{34}$$

Let
$$M_2 = max(\mid a_1 \mid, \mid a_2 \mid, ..., \mid a_N - 1 \mid) \tag{35}$$

Let $M = max(M_1, M_2)$. Then from (34) and (35), we get $\mid a_n \mid < M$, for all $n = 1, 2, 3, ...$

Hence it follows that $\{a_n\}_{n=1}^{\infty}$ is bounded. Thus we have proved that a cauchy sequence is always bounded. Therefore, $\{a_n\}_{n=1}^{infty}$ is bounded sequence and it must contain a convergent subsequence. Let $\{a_{n_k}\}_{n=1}^{\infty}$ be that convergent subsequence, say $a_{n_k} \to a$; i.e. a_{n_k} converges to a.

Therefore, given any $\epsilon > 0$ howsoever small, we can find a positive integer U such that for all $n_k \geq U$

$$\mid a_{n_k} - a \mid < \epsilon \tag{36}$$

Now $\mid a_n - a \mid$

$$=\mid a_n - a_{n_k} + a_{n_k} - a \mid$$

$$\leq \mid a_n - a_{n_k} \mid + \mid a_{n_k} - a \mid \text{ (By triangle inequality)}$$

Therefore, $\mid a_n - a \mid < \epsilon + \epsilon$ (by 36) and the fact that $\{a_n\}_{n=1}^{\infty}$ is a Cauchy sequence, and therefore, $\mid a_n - a_{n_k} \mid < \epsilon$, for all $n, n_k \geq U$)

Therefore, $\mid a_n - a \mid < 2\epsilon$, for all $n \geq U$ which implies that $a_n \to a$

Hence it follows that $\{a_n\}_{n=1}^{\infty}$ is convergent.

Thus the theorem is completely proved. $\square$

Remark 2.26. We have proved that if $\{a_n\}_{n=1}^{\infty}$ is a Cauchy sequence then it is always bounded.

Theorem 2.27. *Prove that a convergent sequence is always bounded.*

Proof. Theorem (2.25) implies that if $\{a_n\}_{n=1}^{\infty}$ is convergent, then it is Cauchy sequence. Also by the same Theorem we know that every cauchy sequence is bounded.

Thus if $\{a_n\}_{n=1}^{\infty}$ is convergent it is bounded. $\qquad\qquad\square$

Exercise 2.28. If $lim_{n\to\infty} a_n = a$, then prove that

$$lim_{n\to\infty} A_n = a$$

where $A_n = \frac{a_1 + a_2 + ... + a_n}{n}$.

Give an example to show that converse is not true.

Solution: Suppose $lim_{n\to\infty} a_n = a$. Let $a_n = a + t_n$, $n = 1, 2, 3, ...$ Then $lim_{n\to\infty} a_n = lim_{n\to\infty} a + lim_{n\to\infty} t_n$, which implies that $a = a + lim_{n\to\infty} t_n$

i.e.
$$\lim_{n\to\infty} t_n = 0 \qquad\qquad (37)$$

Now $A_n = \frac{a_1 + a_2 + ... + a_n}{n}$

or, $A_n = \frac{a + t_1 + a + t_2 + ... + a + t_n}{n}$ (because $a_n = a + t_n$, $n = 1, 2, ..., n$)

or, $A_n = \frac{n(a) + t_1 + t_2 + ... + t_n}{n}$

or, $A_n = a + \frac{t_1 + t_2 + ... + t_n}{n}$

or, $A_n = a + T_n$, where $T_n = \frac{t_1 + t_2 + ... + t_n}{n}$

In order to prove the result we now prove that $lim T_n = 0$.

By (37) $lim_{n\to 0}t_n = 0$, therefore, given $\epsilon > 0$ howsoever small we can find a positive integer N such that $\mid t_n - 0 \mid < \epsilon$, for all $n \geq N$

or, $\mid t_n \mid < \epsilon$, for all $n \geq N$.

Therefore,

$$\mid t_N \mid < \epsilon, \mid t_{n+1} \mid < \epsilon \tag{38}$$

Now $T_n = \frac{t_1+t_2+...+t_{N-1}+t_N+t_{N+1}+...+t_n}{n}$,

therefore, $\mid T_n \mid = \frac{|t_1+t_2+...+t_{N-1}+t_N+t_{N+1}+...+t_n|}{n}$

or, $\mid T_n \mid \leq \frac{1}{n}\{\mid t_1 + t_2 + ... + t_{N-1} \mid + \mid t_N + t_{N+1} + ... + t_n \mid\}$

or, $\mid T_n \mid < \frac{1}{n} \mid t_1 + t_2 + ... + t_{N-1} \mid + \epsilon(\frac{n-N+1}{n})$

or, $\mid T_n \mid < \frac{1}{n} \mid t_1 + t_2 + ... + t_{N-1} \mid + \frac{\epsilon n}{n} - \frac{\epsilon N}{n} + \frac{\epsilon}{n}$

or, $\mid T_n \mid < \frac{1}{n} \mid t_1 + t_2 + ... + t_{N-1} \mid + \epsilon - \frac{\epsilon N}{n} + \frac{\epsilon}{n}$

Therefore, $\mid T_n \mid < \mid \frac{t_1+t_2+...+t_{N-1}}{n} \mid + \epsilon + \frac{\epsilon}{n}$

or, $\mid T_n \mid < \mid \frac{t_1+t_2+...+t_{N-1}}{n} \mid + \epsilon + \epsilon$ (because $\frac{\epsilon}{n} < \epsilon$)

or, $\mid T_n \mid < \mid \frac{t_1+t_2+...+t_{N-1}}{n} \mid + 2\epsilon$

or, $\mid T_n \mid < \frac{k}{n} + 2\epsilon$ (where $k = t_1 + t_2 + ... + t_{N-1}$ and thus K does not depend on n)

or, $T_n < \epsilon + 2\epsilon$, for all $n \geq N'$ where $N' = [\frac{k}{\epsilon}] + 1$ (As $\frac{k}{n} < \epsilon$) if $n > \frac{k}{\epsilon}$, choose $N' = [\frac{k}{\epsilon}] + 1$. Then $\mid T_n \mid < 3\epsilon$, for all $n \geq N'$, which implies that $lim_{n\to\infty}T_n = 0$. But $A_n = a + T_n$, therefore, $lim_{n\to\infty}A_n = lim_{n\to\infty}a + lim_{n\to\infty}T_n = a + 0$

Therefore, $lim_{n\to\infty}A_n = a$.

Converse of above result need not to be true:

Consider $a_n = (-1)^n, n = 1, 2, 3, ...,$

$A_n = \frac{a_1+a_2+......a_n}{n}$,

$A_n = 0$ if n is even and $A_n = \frac{-1}{n}$ if n is odd, therefore, it follows that $lim_{n\to\infty}A_n = 0$. But on the other hand $lim_{n\to\infty}(-1)^n$ does not exist. Hence the result.

Corollary 2.29. *If $a_n > 0$, $a > 0$ and $lim_{n\to\infty}a_n = a$, then $lim_{n\to\infty}G_n = a$, where $G_n = (a_1.a_2...a_n)^{\frac{1}{n}}$.*

Proof. It is given that $a_n > 0$, $a > 0$ and $lim_{n\to\infty}a_n = a$; i.e. $a_n \to a$

or, $loga_n \to loga$. Now by Exercise (2.28), it follows that

$$\frac{loga_1 + loga_2 + ... + loga_n}{n} \to log(a)$$

or, $\frac{1}{n}log(a_1a_2...a_n) \to log(a)$

or, $log(a_1a_2...a_n)^{\frac{1}{n}} \to log(a)$

or, $(a_1a_2...a_n)^{\frac{1}{n}} \to a$.

Therefore, it follows that $lim_{n\to\infty}G_n = a$, where $G_n = (a_1a_2...a_n)^{\frac{1}{n}}$.

$\square$

Exercise 2.30. Prove that $lim_{n\to\infty}n^{\frac{1}{n}} = 1$.

Proof. (Recall exercise 2.23) Consider the sequence $\{a_n\}_{n=1}^{\infty}$, where $a_n = 1$, $n = 1$ and $a_n = \frac{n}{n-1}$, $n > 1$. Then $a_n > 0$, for all n. Also $lim_{n\to\infty}a_n = 1 > 0$, therefore, by above corollary it follows that $lim_{n\to\infty}(a_1.a_2...a_n)^{\frac{1}{n}} = 1$

or $lim_{n\to\infty}(1.\frac{2}{1}.\frac{3}{2}.\frac{4}{3}...\frac{n-1}{n-2}.\frac{n}{n-1})^{\frac{1}{n}} = 1$

Therefore, $lim_{n\to\infty}n^{\frac{1}{n}} = 1$.

$\square$

Definition 2.31. (Limit superior and limit inferior): Consider a sequence $\{a_n\}_{n=1}^{\infty}$. Then limit superior of $\{a_n\}_{n=1}^{\infty}$ is denoted by $\overline{lim}_{n\to\infty}a_n$, and is defined as

$\overline{lim}_{n\to\infty}a_n = a$ if given $\epsilon > 0$ how so ever small, there exists a positive integer N such that

$a_n < a + \epsilon$ for all $n \geq N$ and

$a_n > a - \epsilon$ for an infinite number of values of n.

Also limit inferior of $\{a_n\}_{n=1}^{\infty}$ is denoted by $\underline{lim}_{n\to\infty}a_n$, and is defined as

$\underline{lim}_{n\to\infty}a_n = b$ if given $\epsilon > 0$ how so ever small, there exists a positive integer N such that

$a_n < b - \epsilon$ for all $n \geq N$ and

$a_n > b + \epsilon$ for an infinite number of values of n.

Theorem 2.32. *Prove that* $\overline{lim}_{n\to\infty}a_n$ *is unique.*

Proof. Suppose $\overline{lim}_{n\to\infty}a_n = a$. We have to prove that this limit is unique. If possible suppose $\overline{lim}_{n\to\infty}a_n = b$, where $a \neq b$, say $a < b$. We choose ϵ such that $0 < \epsilon < \frac{b-a}{2}$. Now $a < b$, therefore $b - a > 0$, or $\epsilon < \frac{b-a}{2}$, or $2\epsilon < b - a$; i.e.

$$a + \epsilon < b - \epsilon \tag{39}$$

Since $\overline{lim}_{n\to\infty}a_n = a$, therefore, for the above choice of ϵ we can find a positive integer N_1 such that

$a_n < a + \epsilon$, for all $n \geq N_1$

$a_n > a - \epsilon$, for an infinite number of values of n.

Again since $\overline{lim}_{n\to\infty}a_n = b$, therefore, for the above choice of ϵ we can find a positive integer N_2 such that

$a_n < b + \epsilon$, for all $n \geq N_2$

$a_n > b - \epsilon$, for an infinite number of values of n.

Again since $\overline{lim}_{n\to\infty}a_n = b$

Let $N = \max(N_1, N_2)$. Then from the above set of equations we get for all $n \geq N$

$$a_n < a + \epsilon \tag{40}$$

and for an infinite number of value of n

$$a_n > a - \epsilon \tag{41}$$

Also, for all $n \geq N$

$$a_n < b + \epsilon \tag{42}$$

and for an infinite number of value of n

$$a_n > b - \epsilon \tag{43}$$

. Now from (40) it follows that a finite number of $a'_n s$ lie on right side of $a + \epsilon$ but by (43), $a_n > b - \epsilon$, for an infinite number of values of n.

Therefore, by (39) $a_n > a + \epsilon$, for an infinite number of values of n which contradicts the above.

Hence our supposition that $a < b$ must be wrong. Therefore, $a \not< b$. Similarly we can prove that $a \not> b$, and so it follows that $a = b$.

Hence it follows that $\overline{lim}_{n \to \infty} a_n$ is unique. $\qquad\qquad\square$

Theorem 2.33. *Prove that $\underline{lim}_{n \to \infty} a_n$ is unique.*

Proof. Suppose $\underline{lim}_{n \to \infty} a_n = a$. We have to prove that this limit is unique. If possible suppose $\underline{lim}_{n \to \infty} a_n = b$, where $a \neq b$.

Suppose $a > b$. We choose ϵ such that $0 < \epsilon < \frac{a-b}{2}$. Now $a < b$, therefore $a - b > 0$, or $\epsilon < \frac{a-b}{2}$, or $2\epsilon < a - b$; i.e.

$$b + \epsilon < a - \epsilon \tag{44}$$

Now since $\underline{lim}_{n \to \infty} a_n = a$, therefore, for the above choice of ϵ we can find a positive integer N_1 such that

$a_n > a - \epsilon$, for all $n \geq N_1$ and

$a_n < a + \epsilon$, for an infinite number of values of n.

Again since $\underline{\lim}_{n \to \infty} a_n = b$, therefore, for the above choice of ϵ, we can find a positive integer N_2 such that

$a_n > b - \epsilon$, for all $n \geq N_2$ and

$a_n < b + \epsilon$, for an infinite number of values of n.

Let $N = \max(N_1, N_2)$, then from the above set of equations we get for all $n \geq N$

$$a_n > a - \epsilon \tag{45}$$

and for an infinite number of value of n

$$a_n < a + \epsilon \tag{46}$$

Also for all $n \geq N$

$$a_n > b - \epsilon \tag{47}$$

and for an infinite number of value of n

$$a_n < b + \epsilon \tag{48}$$

Now from (45) it follows that a finite number of $a_n's$ lie on left side of $a - \epsilon$, but by (48)

$a_n < b + \epsilon$, for an infinite number of values of n.

Therefore, by (44), $a_n < b + \epsilon < a - \epsilon$ i.e. $a_n < a - \epsilon$ for an infinite number of values of n, which means that infinite number of $a_n's$ lie on the left of $a - \epsilon$, which contradicts the above.

Hence our supposition that $a > b$ must be wrong. Therefore $a \not> b$. Similarly we can prove that $a \not< b$, and therefore, it follows that $a = b$.

Hence it follows that $\underline{\lim}_{n \to \infty} a_n$ is unique. $\qquad\square$

Theorem 2.34. *Prove that $\underline{\lim}_{n \to \infty} a_n \leq \overline{\lim}_{n \to \infty} a_n$.*

Proof. We have to prove that $\underline{lim}_{n\to\infty}a_n \leq \overline{lim}_{n\to\infty}a_n$.

If possible suppose that $\underline{lim}_{n\to\infty}a_n > \overline{lim}_{n\to\infty}a_n$.

Suppose $\underline{lim}_{n\to\infty}a_n = \lambda$ and $\overline{lim}_{n\to\infty}a_n = \mu$. Then $\lambda - \mu > 0$. We choose ϵ such that

$$0 < \epsilon < \frac{\lambda-\mu}{2}$$

or, $\epsilon < \frac{\lambda-\mu}{2}$

or, $2\epsilon < \lambda - \mu$, i.e.

$$\mu + \epsilon < \lambda - \epsilon \tag{49}$$

Since $\underline{lim}_{n\to\infty}a_n = \lambda$, therefore, for the above choice of ϵ, we can find a positive integer N_1 such that for all $n \geq N_1$

$$a_n > \lambda - \epsilon \tag{50}$$

and for an infinite number of values of n

$$a_n < \lambda + \epsilon \tag{51}$$

Again since $\overline{lim}_{n\to\infty}a_n = \mu$, therefore, for the above choice of ϵ, we can find a positive integer N_2 such that for all $n \geq N_2$

$$a_n < \mu + \epsilon, \tag{52}$$

and for an infinite number of values of n

$$a_n > \mu - \epsilon \tag{53}$$

Let $N = \max (N_1, N_2)$. Then from the above set of equations we get for all $n \geq N$

$$a_n > \lambda - \epsilon \tag{54}$$

and for an infinite number of values of n

$$a_n < \lambda + \epsilon \tag{55}$$

Also for all $n \geq N$

$$a_n < \mu + \epsilon \tag{56}$$

and for an infinite number of values of n

$$a_n > \mu - \epsilon \tag{57}$$

Now from (54), it follows that a only finite number of $a'_n s$ lie on left of $\lambda - \epsilon$ but by (56)

$a_n < \mu + \epsilon$, for an infinite number of values of n. Therefore, by (49)

$a_n < \mu + \epsilon < \lambda - \epsilon$, for all n $\geq N$;

i.e. $a_n < \lambda - \epsilon$, for all $n \geq N$ from which it follows that an infinite number of $a'_n s$ lie on the left of $\lambda - \epsilon$ which is a contradiction.

Hence our supposition that $\underline{lim}_{n \to \infty} a_n > \overline{lim}_{n \to \infty} a_n$ must be wrong and therefore it follows that

$$\underline{lim}_{n \to \infty} a_n \leq \overline{lim}_{n \to \infty} a_n. \qquad \square$$

Theorem 2.35. *Prove that $\{a_n\}_{n=1}^{\infty}$ is convergent if and only if $\underline{lim}_{n \to \infty} a_n = \overline{lim}_{n \to \infty} a_n$.*

Proof. Suppose $\underline{lim}_{n \to \infty} a_n = \overline{lim}_{n \to \infty} a_n = a$ (say)

Since $lim_{n \to \infty} a_n = a$, therefore, given $\epsilon > 0$, we can find a positive integer N such that for all $n \geq N$,

$$a_n < a + \epsilon \tag{58}$$

and for an infinite number of values of n

$$a_n > a - \epsilon \tag{59}$$

Also, $\overline{lim}_{n \to \infty} a_n = a$, therefore, given $\epsilon > 0$, We can find a positive integer N such that for all $n \geq N$

$$a_n > a - \epsilon \tag{60}$$

and for an infinite number of values of n

$$a_n < a + \epsilon \tag{61}$$

Therefore, from (58) and (59), We get

$$a - \epsilon < a_n < a + \epsilon, \text{ for all } n \geq N$$

or

$$-\epsilon < a_n - a < \epsilon, \text{ for all } n \geq N.$$

Therefore we have $\mid a_n - a \mid < \epsilon$, for all $n \geq N$, which implies that $lim_{n \to \infty} a_n = a$. So $\{a_n\}_{n=1}^{\infty}$ is convergent.

Conversely suppose that $\{a_n\}_{n=1}^{\infty}$ is convergent. Therefore, $lim_{n \to \infty} a_n = a$ (say). So for a given $\epsilon > 0$, we can find a positive integer N such that for all $n \geq N \mid a_n - a \mid < \epsilon$ i.e. $-\epsilon < a_n - a < \epsilon$, for all $n \geq N$

or, $a - \epsilon < a_n < a + \epsilon$, for all $n \geq N$

Now $a_n < a + \epsilon$, for all $n \geq N$ implies that

$$\overline{lim}_{n \to \infty} a_n < a + \epsilon \tag{62}$$

Also $a_n > a - \epsilon$, for all $n \geq N$, which implies that $\underline{lim}_{n \to \infty} a_n > a - \epsilon$

or,

$$-\underline{lim}_{n \to \infty} a_n < -a + \epsilon \tag{63}$$

By adding (62) and (63), we get

$$0 \leq \overline{lim}_{n \to \infty} a_n - \underline{lim}_{n \to \infty} a_n < 2\epsilon$$

Since ϵ is arbitrary positive number, therefore, $\epsilon \to 0$ and we have

$$0 \leq \overline{lim}_{n \to \infty} a_n - \underline{lim}_{n \to \infty} a_n \leq 0$$

This implies that $\overline{lim}_{n\to\infty}a_n - \underline{lim}_{n\to\infty}a_n = 0$; i.e. $\overline{lim}_{n\to\infty}a_n = \underline{lim}_{n\to\infty}a_n$.

$\square$

Remark 2.36. : We now give another proof of the fact that a cauchy sequence is convergent. Suppose $\{a_n\}_{n=1}^{\infty}$ is a cauchy sequence , therefore, given $\epsilon > 0$ we can find a positive integer N such that $|a_n - a_m < \epsilon$, for all $n, m \geq N$.

In particular for $m = N$ we get

$|a_n - a_N| < \epsilon$, for all $n \geq N$

or for all $n \geq N$

$$-\epsilon < a_n - a_N < \epsilon. \tag{64}$$

Therefore, $a_n - a_N < \epsilon$, for all $n \geq N$

$a_n < a_N + \epsilon$, for all $n \geq N$,

i.e.

$$\overline{lim}_{n\to\infty}a_n \leq a_N + \epsilon \tag{65}$$

Also from (65),

$-\epsilon < a_n - a_N$

or, $a_N - \epsilon < a_n$

or, $a_n > a_N - \epsilon$

$\underline{lim}_{n\to\infty}a_n \geq a_N - \epsilon$

or,

$$-\underline{lim}_{n\to\infty}a_n \leq -a_N + \epsilon \tag{66}$$

Adding (65) and (66), we get

$$\overline{lim}_{n\to\infty}a_n - -\underline{lim}_{n\to\infty}a_n \leq 2\epsilon. \tag{67}$$

Since $\overline{lim}_{n\to\infty}a_n \leq -\underline{lim}_{n\to\infty}a_n$,

therefore, $\overline{lim}_{n\to\infty}a_n - \underline{lim}_{n\to\infty}a_n \geq 0$.

Now from (67), we get that,

$0 \leq \overline{lim}_{n\to\infty}a_n - \underline{lim}_{n\to\infty}a_n < 2\epsilon.$

Letting $\epsilon \to 0$(since ϵ is arbitrary)

$0 \leq \overline{lim}_{n\to\infty}a_n - \underline{lim}_{n\to\infty}a_n < 0$

Hence $\overline{lim}_{n\to\infty}a_n - \underline{lim}_{n\to\infty}a_n = 0$

or, $\overline{lim}_{n\to\infty}a_n = \underline{lim}_{n\to\infty}a_n$

which proves that$\{a_n\}_{n=1}^{\infty}$ is convergent.(by above theorem (2.35)).

Exercise 2.37. If $s_n = 1 + 1 + \frac{1}{2!} + \frac{1}{3!} + \frac{1}{4!} + ... + \frac{1}{n!}$
then prove that $lim_{n\to\infty}s_n$ exists and is equal to e.

Solution: Before proving this result we mention that

$$lim_{n\to\infty}(1 + \frac{1}{n})^n = e \tag{68}$$

Now $s_n = 1 + 1 + \frac{1}{2!} + \frac{1}{3!} + \frac{1}{4!} + ... + \frac{1}{n!}.$

So $s_{n+1} = \{1 + 1 + \frac{1}{2!} + \frac{1}{3!} + \frac{1}{4!} + ... + \frac{1}{n!}\} + \frac{1}{(n+1)!} = s_n + \frac{1}{(n+1)!}.$

Therefore, $s_{n+1} > s_n$ (because $\frac{1}{(n+1)!} > 0$)

i.e. $\{s_n\}_{n=1}^{\infty}$ is monotonic increasing,

Now $s_n = 1 + 1 + \frac{1}{2!} + \frac{1}{3!} + \frac{1}{4!} + ... + \frac{1}{n!}$

or, $s_n = 2 + \frac{1}{2!} + \frac{1}{3!} + \frac{1}{4!} + ... + \frac{1}{n!}$

or, $s_n < 2 + \frac{1}{2} + \frac{1}{2^2} + \frac{1}{2^3} + ... + \frac{1}{2^n}$ [because $\frac{1}{3!} < \frac{1}{2^2}, \frac{1}{4!} < \frac{1}{2^3}...$]

or, $s_n < 2 + \frac{1}{2} + \frac{1}{2^2} + \frac{1}{2^3} + ...$

or, $s_n < 2 + \frac{\frac{1}{2}}{1 - \frac{1}{2}} = 2 + \frac{\frac{1}{2}}{\frac{1}{2}} = 2 + 1 = 3,$

which implies $s_n < 3$

therefore, it follows that $\{a_n\}_{n=1}^{\infty}$ is bounded above.

Thus $\{a_n\}_{n=1}^{\infty}$ is monotonic increasing and bounded above and therefore it is convergent.

Hence $lim_{n \to \infty} s_n$ exists.

Suppose $t_n = (1 + \frac{1}{n})^n$

Then $lim_{n \to \infty} t_n = e$ using (68)

Now, $t_n = (1 + \frac{1}{n})^n = 1 + 1 + \frac{n(n-1)}{2!} \frac{1}{n^2} + \frac{n(n-1)(n-2)}{3!} \frac{1}{n^3} + ... + \frac{n(n-1)(n-2)...(n-(n-1))}{n!} \frac{1}{n^n}$

or, $t_n = 1 + 1 + (1 - \frac{1}{n})\frac{1}{2!} + (1 - \frac{1}{n})(1 - \frac{2}{n})\frac{1}{3!} + ... + (1 - \frac{1}{n})(1 - \frac{2}{n})...(1 - \frac{n-1}{n})\frac{1}{n!}$

or, $t_n \leq 1 + 1 + \frac{1}{2!} + \frac{1}{3!} + \frac{1}{4!} + ... + \frac{1}{n!} = s_n$

i.e. $t_n \leq s_n$

$\Rightarrow lim_{n \to \infty} t_n \leq lim_{n \to \infty} s_n$

$\Rightarrow e \leq lim_{n \to \infty} s_n$ (because $lim_{n \to \infty} t_n = e$)

Suppose $lim_{n \to \infty} s_n = s$, then from above

$$e \leq s. \tag{69}$$

Now suppose m is a positive integer less than n.

Now $t_n = (1 + \frac{1}{n})^n$

$= 1 + 1 + (1 - \frac{1}{n})\frac{1}{2!} + ... + (1 - \frac{1}{n})(1 - \frac{2}{n})...(1 - \frac{m-1}{n})\frac{1}{m!} + ... + (1 - \frac{1}{n})(1 - \frac{2}{n})...(1 - \frac{n-1}{n})\frac{1}{n!}$

or, $t_n > 1 + 1 + (1 - \frac{1}{n})\frac{1}{2!} + (1 - \frac{1}{n})(1 - \frac{2}{n})\frac{1}{3!} + ... + (1 - \frac{1}{n})(1 - \frac{2}{n})...(1 -$

$\frac{m-1}{n})\frac{1}{m!}$

Keeping m fixed and letting n tend to ∞, we get

$$lim_{n\to\infty}t_n \geq 1 + 1 + \tfrac{1}{2!} + \tfrac{1}{3!} + \tfrac{1}{4!} + ... + \tfrac{1}{m!} = s_m$$

i.e. $lim_{n\to\infty}t_n \geq s_m$, which implies that $e \geq s_m$.

Since m is arbitrary, therefore, letting m tend to ∞, we get $e \geq lim_{m\to\infty}S_m$

or,

$$e \geq s \tag{70}$$

From (69) and (70) it follows that $e = s$.

Therefore, $lim_{n\to\infty}s_n = e$

Exercise 2.38. Prove that e is irrational.

Solution: Before proving this result we will first prove the previous result i.e. If $s_n = 1 + 1 + \tfrac{1}{2!} + \tfrac{1}{3!} + \tfrac{1}{4!} + ... + \tfrac{1}{n!}$,

then prove that $lim_{n\to\infty}s_n$ exists and is equal to e.

Now we have $lim_{n\to\infty}s_n = e$.

Therefore, $e = [1 + 1 + \tfrac{1}{2!} + \tfrac{1}{3!} + ... + \tfrac{1}{n!}] + \tfrac{1}{(n+1)!} + \tfrac{1}{(n+2)!} + ...$

or, $e = s_n + \tfrac{1}{(n+1)!} + \tfrac{1}{(n+2)!} + \tfrac{1}{(n+3)!} + ...$

or, $0 < e - s_n = \tfrac{1}{(n+1)!} + \tfrac{1}{(n+2)(n+1)!} + \tfrac{1}{(n+3)(n+2)(n+1)!} + ...$

or, $0 < e - s_n = \tfrac{1}{(n+1)!}[1 + \tfrac{1}{n+2} + \tfrac{1}{(n+2)(n+3)} + ...]$.

Thus $0 < e - s_n = \tfrac{1}{(n+1)!}[1 + \tfrac{1}{n+2} + \tfrac{1}{(n+2)(n+3)} + ...]$

or, $0 < e - s_n = \tfrac{1}{(n+1)!}[1 + \tfrac{1}{n+1} + \tfrac{1}{(n+1)^2} + ...]$ (because $\tfrac{1}{n+2} < \tfrac{1}{n+1}$)

or, $0 < e - s_n = \frac{1}{(n+1)!}\left[\frac{1}{1-\frac{1}{n+1}}\right]$

or, $0 < e - s_n = \frac{1}{(n+1)!}\frac{n+1}{n} = \frac{1}{(n+1)n!}\frac{n+1}{n} = \frac{1}{n.n!}$

Therefore,

$$0 < e - s_n < \frac{1}{n.n!} \tag{71}$$

Now we have to prove that e is irrational. If possible suppose e is rational say $e = \frac{p}{q}$, $q \neq 0$, $q \geq 1$.

Therefore,

$$\frac{1}{q} \leq 1 \tag{72}$$

From (71), we get $0 < \frac{p}{q} - s_n < \frac{1}{n.n!}$.

Putting $n = q$, we get

$0 < \frac{p}{q} - s_q < \frac{1}{q.q!}$

or, $0 < \frac{p}{q} - \{1 + 1 + \frac{1}{2!} + \frac{1}{3!} + ... + \frac{1}{q!}\} < \frac{1}{q.q!}$

Multiplying by $q!$ we get

$0 < \frac{p}{q}.q! - \{q! + q! + \frac{q!}{2!} + \frac{q!}{3!} + ... + 1\} < \frac{1}{q} \leq 1$ (by using (72))

i.e. $0 < integer - integer < 1$ [because $\frac{p}{q}.q!, q!, \frac{q!}{2!}, \frac{q!}{3!}, ..., \frac{q!}{q!}$ are all integers]

or, $0 < integer < 1$

which is clearly a contradiction because there is no integer which lies between 0 and 1. Hence our supposition that e is rational must be wrong.

It follows that e is irrational.

Exercise 2.39. If $\{a_n\}_{n=1}^{\infty}$, $a_n \geq 0$ is such that $lim_{n \to \infty} = a$, then $a \geq 0$.

Solution: $\{a_n\}_{n=1}^{\infty}$ is given to be such that $a_n \geq 0$ and $lim_{n\to\infty} = a$. We have to prove that $a \geq 0$.

If possible suppose the result is false, then $a < 0$, i.e. $-a > 0$, we choose $\epsilon > 0$ such that $\epsilon < -a$ or

$$a + \epsilon < 0 \tag{73}$$

Since it is given that $lim_{n\to\infty} = a$, therefore, for above choice of ϵ we can find a positive integer N such that

$|a_n - a| < \epsilon$, for all $n \geq N$ or

$-\epsilon < a_n - a < \epsilon$, for all $n \leq N$

or $a_n - a < \epsilon$, for all $n \geq N$

or $a_n < a + \epsilon$, for all $n \geq N$

or $a_n < a + \epsilon < 0$, for all $n \geq N$, using (73)

or $a_n < 0$ which is clearly a contradiction because $a_n \geq 0$, for all $n \geq N$

Therefore, our supposition $a < 0$ must be wrong.

Hence, it follows that $a \geq 0$.

Exercise 2.40. Let $0 < a < b$ and

$$s_1 = a, \ s_2 = \sqrt{\frac{ab^2 + s_1^2}{a+1}}, \ \ldots, \ s_{n+1} = \sqrt{\frac{ab^2 + s_n^2}{a+1}}, \ldots$$

Prove that $lim_{n\to\infty} s_n = b$.

Solution: Before proving this result we prove the previous result, i.e. if $\{a_n\}_{n=1}^{\infty}$, $a_n \geq 0$ is such that $lim_{n\to\infty} = a$, then $a \geq 0$.

Now, $s_1 = a, \ s_2 = \sqrt{\frac{ab^2 + s_1^2}{a+1}}, \ \ldots, \ s_{n+1} = \sqrt{\frac{ab^2 + s_n^2}{a+1}}, \ldots$

Therefore, $s_{n+1}^2 - s_n^2 = \frac{ab^2+s_n^2}{a+1} - \frac{ab^2+s_{n-1}^2}{a+1} = s_n^2 - s_{n-1}^2$

i.e. $s_{n+1}^2 - s_n^2 = s_n^2 - s_{n-1}^2$. This being a recurrence relation implies that

$$s_{n+1} > s_n \text{ if } s_n > s_{n-1}.$$

By the repeated application of this it follows that

$$s_{n+1} > s_n \text{ if } s_2 > s_1 \text{ if } s_2^2 > s_1^2$$

i.e. $s_{n+1} > s_n$ if $\sqrt{\frac{ab^2+s_1^2}{a+1}} > a$

i.e. $s_{n+1} > s_n$ if $\frac{ab^2+s_1^2}{a+1} > a^2$

i.e. $s_{n+1} > s_n$ if $\frac{ab^2+a^2}{a+1} > a^2$

i.e. $s_{n+1} > s_n$ if $b^2 > a^2$

i.e. $s_{n+1} > s_n$ if $b > a$ which is true

So, it follows that $s_{n+1} > s_n$.

Therefore, $\{s_n\}_{n=1}^{\infty}$ is monotonic increasing.

We now prove that $\{s_n\}_{n=1}^{\infty}$ is bounded above by using mathematical induction.

$s_1 = a$, and therefore, so $s_1 < b$ (because $0 < a < b$).

Now suppose $s_n < b$, then

$$s_{n+1} = \sqrt{\frac{ab^2+s_n^2}{a+1}}$$

implies that $s_{n+1} < \sqrt{\frac{ab^2+b^2}{a+1}}$

or, $s_{n+1} < \sqrt{\frac{b^2(a+1)}{a+1}} = \sqrt{b^2} = b$

Therefore, $s_{n+1} < b$, and so $s_n < b$ for all n.

Thus $\{s_n\}_{n=1}^{\infty}$ is bounded above.

Therefore, $\{s_n\}_{n=1}^{\infty}$ is monotonic increasing and bounded above and hence it is convergent say $lim_{n\to\infty} s_n = t$.

Now $s_n > 0$, for all n, so by using previous result, we have $t > 0$.

Now $s_{n+1} = \sqrt{\frac{ab^2 + s_n^2}{a+1}}$

or, $s_{n+1}^2 = \frac{ab^2 + s_n^2}{a+1}$,

Letting $n \to \infty$ we get

$t^2 = \frac{ab^2 + t^2}{a+1}$

i.e. $at^2 + t^2 = ab^2 + t^2$

or, $at^2 = ab^2$

or, $t^2 = b^2$

Now since $t > 0, b > 0$, we have

$t = b$.

Hence the result.

EXERCISES

(1) Show that

    (a) $lim_n \frac{2n-3}{n+1} = 2$

    (b) $lim_n \frac{1+2+...+n}{n^2} = \frac{1}{2}$

    (c) $lim_n \frac{1+3+5+...+(2n-1)}{n^2} = 1$

(2) If $lim_n x_n = x$ and $lim_n y_n = y$, then prove that

$$lim_n \frac{x_1 y_n + x_2 y_{n-1} + ... + x_n y_1}{n} = xy$$

(3) Evaluate the following:

    (a) $lim_n \frac{n}{(n!)^{\frac{1}{n}}}$

    (b) $lim_n \left(\frac{2n!}{n!n!}\right)^{\frac{1}{n}}$

    (c) $lim_n \frac{n!}{n^n}$

(4) Define the lim Sup and lim Inf of a sequence. Determine lim Sup and lim Inf of the following sequences:

    (a) $\{(-1)^n n\}$

    (b) $\{1 - n^{(-1)^n}\}$

    (c) $\{\frac{2n}{3n+1}\}$

(5) Prove that for any sequence $\{x_n\}$, $x_n > 0$ for all n,

$$\lim \inf \frac{x_{n+1}}{x_n} \leq \lim \inf \sqrt{x_n} \leq \lim \sup \frac{x_{n+1}}{x_n}.$$

(6) If $a_n = \sin n\frac{\pi}{2} + \frac{(-1)^n}{n}$, $n \in N$, then show that

$$\underline{lim}a_n = -1; \overline{lim}a_n = 1$$

(7) Show that the sequence $\{x_n\}$ where $x_{n+1} = \sqrt{2x_n}, x_1 = \sqrt{2}$, converges to 2.

(8) Show that the sequence defined by

$$x_{n+1} = \frac{4+3x_n}{3+2x_n}, x_1 = 1, \text{ is convergent with limit } \sqrt{2}$$

(9) Test for convergence the following series:

    (a) $\sum_{n=1}^{\infty} \frac{n+\sqrt{n}}{2n^3-1}$

    (b) $\sum_{n=1}^{\infty} (\sqrt[3]{n^3+1} - n)$

    (c) $\sum_{n=1}^{\infty} \frac{logn}{5n^3-1}$

    (d) $\sum_{n=1}^{\infty} \frac{n(n+1)}{3^{n+1}}$

    (e) $\sum_{n=1}^{\infty} (1 + \frac{1}{n})^{-n^2}$

(10) Test for convergence the following series:

    (a) $\frac{1}{1+2^{-1}} + \frac{2}{1+2^{-2}} + \frac{3}{1+2^{-3}} + \dots \infty$

    (b) $\frac{a}{b} + \frac{a(a+1)}{b(b+1)} + \frac{a(a+1)(a+2)}{a(b+1)(b+2)} + \dots \infty$

    (c) $\frac{1+x}{1!} + \frac{(1+2x)^2}{2!} + \frac{(1+3x)^3}{3!} + \dots \infty$

(11) Discuss the convergence of $\sum_{n=1}^{\infty} \{log(n+1)\}^{-}log(n+1)$.

(12) If $\{x_n\}$ converges to x, Show that $\{y_n\}$ converges also to x where

$$y_n = \frac{1}{2^n}x_1 + \frac{1}{2^{n-1}}x_2 + \dots + \frac{1}{2}x_n, \quad n = 1, 2, \dots$$

(13) Give an example of the series for which $\sum_{n=1}^{\infty} a_n$ converges but $\sum_{n=1}^{\infty} a_n^2$ diverges.

(14) Use cauchy's general principle of convergence to show that following sequences are convergent:

    (a) $\{\frac{n}{n+1}\}$

    (b) $\{1 + \frac{1}{2!} + \frac{1}{3!} + \dots + \frac{1}{n!}\}$

    (c) $\{\frac{(-1)^n}{n}\}$

(15) Let $\{a_n\}$ be a sequence defined by $a_{n+1} = 1 - \sqrt{1-a_n}$, for all $n \geq 1$ and $a_1 < 1$ converges to 0.

(16) Show that the sequence $\{a_n\}$ defined by $a_{n+1} = 1 - \sqrt{1-a_n}$, for all $n \geq 1$ and $a_1 < 1$ converges to 0.

Chapter3

LIMIT AND CONTINUITY

We are familiar with the operations like addition and multiplication etc. We introduce here a typical operation, often called limit operation, performable only on functions. Limit operation play a fundamental role in calculus and most of the concepts here require this operation.

Limit of a function may not exist in all cases, so it should be noted carefully when does limit exist and under what conditions. The notion of limits in respect of functions forms the base for the study of continuity and differentiability etc.

The study of continuity of functions is the most important aspect of analysis and is based on the notion of limit.

Definition 3.1. Limit of a function: suppose a function $f(x)$ is defined in open interval (a, b) except possibly at α, then $f(x)$ converges (tends) to a limit l as $n \to \alpha$ if given $\varepsilon > 0$, we can find a $\delta > 0$ such that

$$|f(x) - l| < \epsilon \text{ for } 0 < |x - \alpha| < \delta,$$

and we write

$$f(x) \to l \text{ as } x \to \alpha \text{ or } \lim_{x \to \alpha} f(x) = l.$$

Definition 3.2. Continuity: A function $f(x)$ of a real variable x is said to be continuous at $x = a$ if $f(x)$ is defined at a and

$$\lim_{h \to 0} f(a - h) = f(a) = \lim_{h \to 0} f(a + h).$$

If f is not continuous at a, then f is said to be discontinues at a. Thus a function can be discontinuous at a if it is not defined at a or

$\lim_{h \to 0} f(a - h) = \lim_{h \to 0} f(a + h)$, but their common value is not equal to $f(a)$ or

$\lim_{h \to 0} f(a - h)$ and $\lim_{h \to 0} f(a + h)$ do not exist or

$\lim_{h \to 0} f(a - h) \neq f(a)$ or $f(a) \neq \lim_{h \to 0} f(a + h)$.

Definition 3.3. Alternate definition of continuity: A function $f(x)$ is said to be continuous at $x = a$ if given an $\epsilon > 0$, there exists $\delta > 0$, such that

$$|f(x) - f(a)| < \epsilon \text{ for } |x - a| < \delta.$$

In other words a function $f(x)$ is said to be continuous at $x = a$ if:

(1) $f(x)$ is defined at a.

(2) $\lim_{x \to a} f(x)$ exists.

(3) $\lim_{x \to a} f(x) = f(a)$; i.e. $f(x) \to f(a)$ as $x \to a$.

Examples:

Examples of functions which are continuous: All trigonometric functions, polynomials, exponentials etc are continuous functions; $sinx$, $tanx$, e^x, $x^2 + 2x + 1$ etc.

(1) $f(x) = 2$, $x \neq 0$.

since $f(x)$ is not defined at 0, therefore, $f(x)$ is discontinuous at 0

(2) $f(x) = 1$, $x \leq 0$............(1)
$\quad f(x) = 2$, $x > 0$............(2)

Here $f(x)$ is defined at 0, as $f(0) = 1$. Also $\lim_{h \to 0} f(0 - h) = \lim_{h \to 0} f(-h) = 1$ by (1) for $-h < 0$.

Similarly $\lim_{h \to 0} f(0 + h) = \lim_{h \to 0} f(h) = 2$ by (2) for $h > 0$.

Thus $\lim_{h \to 0} f(0 - h) \neq \lim_{h \to 0} f(0 + h)$,

therefore, it follows that $f(x)$ is discontinuous at 0.

(3) $f(x) = 1,\ x \geq 0$
$f(x) = 2,\ x < 0$

Now $f(0) = 1$. Also for $h > 0$,

$\lim_{h \to 0} f(0 + h) = \lim_{h \to 0} f(h) = 1$,

and $\lim_{h \to 0} f(0 - h) = \lim_{h \to 0} f(-h) = 2$.

Therefore, $f(0) = \lim_{h \to 0} f(0 - h) \neq \lim_{h \to 0} f(0 + h)$,

which shows that f is discontinuous at point 0.

(4) $f(x) = 1,\ x \neq 0$

$f(x) = 5,\ x = 0$

Here $f(0) = 5$, and for $h > 0$,

$\lim_{h \to 0} f(0+h) = \lim_{h \to 0} f(h) = \lim_{h \to 0} f(0-h) = \lim_{h \to 0} f(-h) = 1$.

Therefore, $\lim_{h \to 0} f(0 + h) = \lim_{h \to 0} f(0 - h) \neq f(0)$,

which shows that $f(x)$ is discontinuous at 0.

(5) $f(x) = 1,\ x$ is rational

$f(x) = 0,\ x$ is irrational

If α is any rational number, then $f(\alpha) = 1$ and for $h > 0$,

$f(\alpha + h) = 1$, if h is rational

$f(\alpha + h) = 0$, if h is irrational.

Therefore, it follows that as $h \to 0$ through positive values, $\lim_{h \to 0} f(\alpha + h)$ does not exist. [because a function has a unique limit].

Similarly $\lim_{h \to 0} f(\alpha - h)$ does not exist. Hence it follows that $f(x)$ is discontinuous at all rational points.

Now suppose β is any irrational number, then $f(\beta) = 0$, and for $h > 0$, $f(\beta + h) = 1$ or 0, depending on h.

So, as $h \to 0$ through positive values $f(\beta + h)$ does not exist (because a function has a unique limit).

Similarly, $f(\beta - h$ does not exist. Therefore, it follows that $f(x)$ is discontinuous at all irrational points.

Thus $f(x)$ is discontinuous at all rational and irrational points and hence discontinuous everywhere.

(6) $f(x) = 1/x,\ x \neq 0$
$f(x) = 1,\ x = 0$

Now for $h > 0$, $\lim_{h \to 0} f(0 + h) = \lim_{h \to 0} f(h) = \lim_{h \to 0} f(1/h) = \infty$

Also, $\lim_{h \to 0} f(0 - h) = \lim_{h \to 0} f(-h) = \lim_{h \to 0} f(-1/h) = -\infty$.

Therefore, $f(0) \neq f(0 + h) \neq f(0 - h)$. Hence, the function is discontinuous at 0.

(7) $f(x) = [x]$.

Let n be any positive integer. Then $f(x) = [n] = n$.

For $h > 0$, $f(n + h) = [n + h] = n$, and

$f(n - h) = [n - h] = n - 1$.

Now $\lim_{h \to 0} f(n + h) = n$, and $\lim_{h \to 0} f(n - h) = [n - h] = n - 1$, which implies that

$\lim_{h \to 0} f(n + h) = f(n) \neq \lim_{h \to 0} f(n - h)$.

Therefore this function is discontinuous at all positive integers.

(8) $f(x) = e^{1/x},\ x \neq 0$

$f(x) = 1,\ x = 0$.

Now, $f(0) = 1$, and for $h > 0$, $f(0 + h) = f(h) = e^{1/h}$, and $f(0 - h) = f(-h) = e^{-1/h}$,

Therefore, $\lim_{h \to 0} f(0 + h) = \lim_{h \to 0} f(h) = \lim_{h \to 0} e^{1/h} = \infty$,

and $\lim_{h \to 0} f(0 - h) = \lim_{h \to 0} f(-h) = \lim_{h \to 0} e^{-1/h} = 0$.

Therefore, $f(0) = f(0 - h) \neq f(0 + h)$, which proves that $f(x)$ is discontinuous at 0.

Definition 3.4. A function f is said to be continuous in $[a, b] = I$ if it is continuous at every point of I.

Theorem 3.5. *suppose a function $f(x)$ is continuous in $[a, b] = I$, then given $\epsilon > 0$, we can divide I into a finite number of subintervals*

$$a = x_0 < x_1 < x_2 < x_3 < ... < x_{k-1} < x_k = b$$

such that $|f(x_j) - f(x_{j+1})| < \epsilon$ for $|x_j - x_{j+1}| < \delta$, where δ is the length of the maximum subinterval.

or

Suppose a function $f(x)$ is continuous in $[a, b] = I$. The given $\epsilon > 0$, we can find a $\delta > 0$ such that the oscillation of $f(x)$ in any close interval, whose length does not exceed δ, is less than ϵ.

Proof. Suppose f is continuous in $[a, b] = I$. We have to prove that given $\epsilon > 0$, $|f(x_j) - f(x_{j+1})| < \epsilon$, whenever x_j and x_{j+1} lie in the same subinterval.

If possible suppose the theorem is false.

Then

$$|f(x_j) - f(x_{j+1})| \geq \epsilon \tag{74}$$

whenever x_j and x_{j+1} lie in the same subinterval.

We bisect I into two close subintervals. Since the theorem is false for I, therefore it will be false for one of close intervals (say I_1) of I. We now bisect I_1 into two close subintervals, since the theorem is false for I_1, the theorem will be false for one of the close subinterval (say I_2) of I_1. We continue this process indefinitely, and at n^{th} step, we bisect close interval I_{n-1} into two close subintervals. Since the theorem is false for I_{n-1}, the theorem will be false for one of the close subinterval (say I_n) of I_{n-1}.

In this way we get a sequence of close intervals $I, I_1, I_2, I_3, ...$ such that

$$I \supseteq I_1 \supseteq I_2 \supseteq I_3 \supseteq \ldots$$

Also $Length(I_n) = \frac{(b-a)}{2^n} \to 0$ as $n \to \infty$. Therefore, by Nested Interval Theorem (3.6), there exists exactly one point x_0 such that $x_0 \in I_n$, for all n. Now $I_n \subset I$, for all n, therefore, $x_0 \in I_n$.

Also $f(x)$ is given to be continues in $[a, b] = I$, therefore, $f(x)$ is continues at x_0 also. So, given $\epsilon > 0$, we can find $\delta > 0$ such that

$$|f(x) - f(x_0)| < \epsilon/2 \tag{75}$$

for $|x - x_0| < \delta$

Now $L(I_n) \to 0$ as $n \to \infty$ and $x_0 \in I_n$, for all n. We choose n so large such that $I_n \subseteq (x_0 - \delta, x_0 + \delta)$.

Now the theorem is false for each I_n implies that it will be false for x_j and x_{j+1} which lie in I_n so that $I_n \subseteq (x_0 - \delta, x_0 + \delta)$. Therefore, from 75 it follows that

$$|f(x_j) - f(x_0)| < \epsilon/2 \tag{76}$$

$$|f(x_{j+1}) - f(x_0)| < \epsilon/2 \tag{77}$$

where x_j and x_{j+1} lie in same interval. Now
$$|f(x_j) - f(x_{j+1})| = |f(x_j) - f(x_0) + f(x_0) - f(x_{j+1})|$$

$$\leq |f(x_j) - f(x_0)| + f(x_0) - f(x_{j+1})|$$

$$< \epsilon/2 + \epsilon/2 = \epsilon$$

or, $|f(x_j) - f(x_{j+1})| < \epsilon$, whenever x_j and x_{j+1} lie in same subinterval, but this is a contradiction to (74).

Hence it follows that theorem must be true. $\qquad\qquad\square$

Theorem 3.6. *If $f(x)$ is continuous in $[a, b] = I$, then it is bounded in I.*

Proof. $f(x)$ is given to be continuous in $[a, b] = I$, therefore given $\epsilon > 0$

we can divide I into a finite number of subintervals such that

$$|f(x') - f(x'')| < \epsilon \tag{78}$$

whenever x' and x'' lie in same interval.

We now divide I into a finite number of subintervals. Let the points of division of I be

$$a = x_0 < x_1 < x_2 < ... < x_{n-1} < x_n = b$$

Let $x \in [x_0, x_1]$. Then

$$|f(x)|$$

$$= |f(x) - f(x_0) + f(x_0)|$$

$$\leq |f(x) - f(x_0)| + |f(x_0)| \text{ [by using triangle inequality]}$$

$$< \epsilon + |f(x_0)| = \epsilon + |f(a)|$$

(because $x_0 = a$ and $f(x)$ is continues in $[a, b] = I$. Therefore it is continues at x_0 and $x \in [x_0, x_1]$. Therefore, if $x \in [x_0, x_1]$, then

$$|f(x)| < \epsilon + |f(a)|. \tag{79}$$

Now using triangle inequality and (79)and the fact that $x \in [x_1, x_2]$ and $f(x)$ is continues in $I]$, we have

$$|f(x)|$$

$$= |f(x) - f(x_1) + f(x_1)|$$

$$\leq |f(x) - f(x_1)| + |f(x_1)|$$

$$< \epsilon + \epsilon + |f(x_0)| = 2\epsilon + |f(a)|.$$

Therefore, if $x \in [x_1, x_2]$, then $|f(x)| < 2\epsilon + |f(a)|$. So, in particular

$$|f(x_2)| < 2\epsilon + |f(a)|. \tag{80}$$

as $x_2 \in [x_1, x_2]$. Similarly, if $x \in [x_2, x_3]$, then $|f(x)| < 3\epsilon + |f(a)|$, and in particular $|f(x_3)| < 3\epsilon + |f(a)|$

Similarly if $x \in [x_{n-1}, x_n]$, then $|f(x)| < n\epsilon + |f(a)|$.

Now

$$n\epsilon + |f(a)| > \epsilon + |f(a)|, 2\epsilon + |f(a)|, 3\epsilon + |f(a)|, ...,(n-1)\epsilon + |f(a)|.$$

Therefore, it follows that

$|f(x)| < n\epsilon + |f(a)|$, for all $x \in I$;

i.e. $|f(x)| < M$, for all $x \in I$, where $M = n\epsilon + |f(a)|$,

which proves that function $f(x)$ is bounded in I. $\qquad\qquad\square$

Theorem 3.7. *Let $f(x)$ be continuous in $[a, b] = I$ so that it is bounded there (say) $m \le f(x) \le M$, for all $x \in I$. Then f assumes m and M in I.*

Proof. Suppose $f(x)$ is continuous in $[a, b] = I$ so that it is bounded in I. Let

$$M = lub_{x \in I}(f(x)), \text{ and } m = glb_{x \in I}(f(x))$$

We have to prove that $f(x) = M$ for some $x \in I$, and $f(y) = m$, for some $y \in I$.

If possible suppose $f(x) \neq M$, for any $x \in I$, then $f(x) < M$, for all $x \in I$; i.e. $M - f(x) > 0$, for all $x \in I$.

Consider $g(x) = \frac{1}{M - f(x)}$, $x \in I$. Then $g(x) > 0$ in I, and is continuous there.

Therefore $g(x)$ must be bounded, say $0 < g(x) \le K$.

Now $g(x) = \frac{1}{M - f(x)}$, $x \in I$, therefore $\frac{1}{M - f(x)} \le \frac{K}{1}$, $x \in I$.

or, $M - f(x) \geq \frac{1}{K}$, $x \in I$

or, $f(x) \leq M - \frac{1}{K}$, $x \in I$,

which is clearly a contradiction because $M = lub_{x \in I}(f(x))$ and $M - \frac{1}{K} < M$.

Therefore our supposition that $f(x) \neq M$, for any $x \in I$ must be wrong. Hence $f(x) = M$ for some $x \in I$.

If possible suppose $f(x) \neq m$, for any $x \in I$, then $f(x) > m$, $x \in I$; i.e. $f(x) - m > 0$, $x \in I$.

Consider $h(x) = \frac{1}{f(x)-m}$, $x \in I$. Then $h(x) > 0$, $x \in I$ and $h(x)$ is continuous in I.

Therefore $h(x)$ is bounded in I, say $0 < h(x) \leq K'$, $x \in I$; i.e. $\frac{1}{f(x)-m} \leq \frac{K'}{1}$, $x \in I$

or, $f(x) - m \geq \frac{1}{K'}$, $x \in I$

or, $f(x) \geq m + \frac{1}{K'}$, $x \in I$,

which is clearly a contradiction because $m = glb_{x \in I}(f(x))$ and $m + \frac{1}{K'} > m$.

Hence our supposition that $f(x) \neq m$ for any $x \in I$ must be wrong.

Therefore $f(x) = m$ for some $x \in I$. Hence the result. $\qquad \square$

Consider $f(x) = \frac{1}{x}$, $x \in (0, 1]$, then $f(x)$ is continuous in $(0, 1]$ (because only point of discontinuity is 0 which does not lie in this interval. But $f(x)$ can be made arbitrarily large by choosing x arbitrary close to 0, which shows that function $f(x)$ is not bounded in $(0, 1]$. Thus the hypothesis that interval must be closed is essential.

Theorem 3.8. *Suppose $f(x)$ is continuous in $[a, b] = I$ and $f(a)$ and $f(b)$ differ in sign. Then there exists $x \in [a, b]$ such that $f(x) = 0$.*

Proof. Suppose $f(x)$ is continuous in $[a, b] = I$ and $f(a)$ and $f(b)$ differ in sign. Since $f(x)$ is continues in $[a, b] = I$, therefore given any $\epsilon > 0$,

$$|f(x') - f(x'')| < \epsilon \tag{81}$$

whenever x' and x'' lie in the same subinterval. Also $f(a)$ and $f(b)$ differ in sign, therefore $f(x)$ must change its sign in at least one subinterval. Let that subinterval contain points x' and x'' such that $f(x') > 0$ and $f(x'') < 0$ so that $f(x') - f(x'') > 0$.

Now $0 < f(x') < f(x') + (-f(x''))$; which implies by (81) that

$$0 < f(x') < f(x') - f(x'') = |f(x') - f(x'')| < \epsilon$$

Therefore,
$$f(x') = |f(x'')| < \epsilon \tag{82}$$

Now consider $g(x) = \frac{1}{f(x)}$, $x \in I$. Then $|g(x)|$ is continuous in I and hence it is bounded there. Now $g(x') = \frac{1}{f(x')}$. Then $|g(x')| > \frac{1}{\epsilon}$ [by (82)].

From this inequality it follows that $|g(x')|$ can be made arbitrarily large by choosing ϵ sufficiently small, which is clearly a contradiction because $g(x)$ is bounded in I as it is continuous there.

Hence our supposition that $f(x)$ does not vanish in $I = [a, b]$ must be wrong.

Therefore $f(x)$ must vanish for at least one $x \in I$; i.e. $f(x) = 0$ for some $x \in I$. Hence the result. $\square$

Corollary 3.9. Intermediate property of continuous functions *If $f(x)$ is continuous in $[a, b]$ and $f(a) < \mu < f(b)$, then there is at least one point c, $a < c < b$ such that $f(c) = \mu$.*

Proof. It is given that $f(x)$ is continuous in $[a, b]$ and $f(a) < \mu < f(b)$. Consider

$$g(x) = f(x) - \mu.$$

Then $g(x)$ is continuous in $[a, b]$. (because μ is constant and is continuous and $f(x)$ is continuous in $[a, b]$).

Now $g(a) = f(a) - \mu < 0$, as $f(a) < \mu$, and $g(b) = f(b) - \mu > 0$, as $f(a) > \mu$.

Therefore, $g(a)$ and $g(b)$ differ in sign, and so by above theorem, it follows that $g(x) = 0$ for some $x \in [a, b]$; i.e. $g(c) = 0$, for some c, $a < c < b$

or, $f(c) - \mu = 0$. Hence the result. $\square$

Remark 3.10. Consider $f(x) = x$, $x \in (0, 1)$. Then $f(x)$ is continuous in $(0, 1)$. But $f(x)$ does not assume the values 1 and 0, where $1 = lub(f(x))$, $x \in (0, 1)$, and $0 = glb(f(x))$, $x \in (0, 1)$

Thus the function $f(x)$ does not assume its bounds. The reason is that the interval is not closed.

Exercise 3.11. Let $f(x) = x$, x is rational; $f(x) = 1 - x$, x irrational, $x \in [0, 1] = I$.

Prove that $f(x)$ is continuous at every point of I other than $\frac{1}{2}$, yet it assumes every value between 0 and 1.

Solution: $f(x) = x$, x is rational; $f(x) = 1 - x$, x irrational, $x \in [0, 1] = I$.

Let α be a rational number. Then

$f(\alpha) = \alpha,$

$f(\alpha \pm h) = \alpha \pm h$; if h is rational, and

$f(\alpha \pm h) = 1 - \alpha \pm h$, if h is irrational.

Therefore,

$\lim_{h \to 0} f(\alpha \pm h) = \alpha$; if h is rational, and

$\lim_{h \to 0} f(\alpha \pm h) = 1 - \alpha$, if h is irrational. Now $\alpha = 1 - \alpha$ implies that $\alpha = \frac{1}{2}$. So if $\alpha = \frac{1}{2}$, then $\lim_{h\to 0} f(\frac{1}{2} \pm h) = \frac{1}{2}f(\frac{1}{2})$. This proves that function $f(x)$ is continues at $\frac{1}{2}$. Now if α is a rational number other than $\frac{1}{2}$, then $2 \neq 1 - \alpha$ and therefore the function $f(x)$ is discontinuous.

Now let β be an irrational number, then $f(\beta) = 1-\beta$. Also $f(\beta\pm h) = \beta \pm h$; if h is rational, and $f(\beta \pm h) = \beta \pm h$ or $(1 - (\beta \pm h))$, if $\beta \pm h$ is rational or irrational respectively. Therefore, $\lim_{h\to 0} f(\beta \pm h)$ does not exist. Hence it follows that function $f(x)$ is discontinuous at β and hence is discontinuous at all irrational points.

Now $f(0) = 0$ and $f(1) = 1$. If α is rational and belongs to $[0, 1]$, then $f(\alpha) = \alpha$. This shows that the function $f(x)$ assumes all rational values in $[0, 1]$. Now let β be an irrational number in $[0, 1]$. Then $0 < \beta < 1$ or $0 < 1 - \beta < 1$ because β is irrational. Therefore $1 - \beta$ is irrational, and so $f(-\beta) = 1 - (1 - \beta) = 1 - 1 + \beta = \beta$. Thus $f(x)$ assumes every irrational value in $[0, 1]$.

Exercise 3.12. If $f(x)$ is continuous and assumes only rational values, then $f(x)$ is constant.

Solution suppose $f(x)$ is continuous and assumes only rational values. Let $x_1 < x_2$. To prove that $f(x)$ is constant , we must prove that $f(x_1) = f(x_2)$.

If possible suppose the result is false.

Then $f(x_1) \neq f(x_2)$, where $f(x_1)$ and $f(x_2)$ are rational say $f(x_1) < f(x_2)$. Now by Irrational Density Theorem (1.19), there exist an irrational number μ such that

$$f(x_1) < \mu < f(x_2) \tag{83}$$

Now consider $[x_1, x_2]$, Then $f(x)$ is continuous in $[x_1, x_2]$ and $f(x_1) < \mu < f(x_2)$. Therefore, by Intermediate Value Theorem (3.9), there exists an element $c \in [x_1, x_2]$ such that $f(c) = \mu$.

Thus $f(c) = \mu$, an irrational number, which is a contradiction, as $f(x)$ assumes only rational values. Therefore, $f(x_1) \not< f(x_2)$. Similarly we can prove that $f(x_1) \not> f(x_2)$

Therefore, it follows that $f(x_1) = f(x_2)$, which shows that $f(x)$ is constant.

Exercise 3.13. Prove that if $f(x)$ is continuous at a point a, then so is $|f(x)|$. Give an example to show that converse is not true.

solution: Let $f(x)$ be continuous. Therefore, given $\epsilon > 0$ we can find $\delta > 0$ such that for $|x - a| < \delta$

$$|f(x) - f(a)| < \epsilon \qquad (84)$$

We know that $|x - y| \geq \big||x| - |y|\big|$,

therefore, $|f(x) - f(a)| \geq \big||f(x)| - |f(a)|\big|$.

So using this in (84), we get that

$$\big||f(x)| - |f(a)|\big| \leq |f(x)| - |f(a)| < \epsilon \text{ for } |x - a| < \delta$$

Therefore, for $|x - a| < \delta$

$$\big||f(x)| - |f(a)|\big| < \epsilon$$

This proves that $|f(x)|$ is continuous at a.

Now consider
$f(x) = 1$, x is rational, and
$f(x) = -1$, x is irrational.
Then $|f(x)| = 1$, for all x, which is continuous everywhere.

Now let α be a rational number. Then $f(\alpha) = 1$, and $f(\alpha \pm h) = 1$ or -1 depending on h. Therefore, $lim_{h \to 0} f(\alpha \pm h)$ does not exist. Therefore, $f(x)$ is discontinuous at all rational points.

Similarly, it can be seen that $f(x)$ is discontinuous at all irrational points also. Therefore it follows that $f(x)$ is discontinuous everywhere.

Exercise 3.14. If $f(x)$ is continuous and assumes only irrational values, then $f(x)$ is constant.

Solution: Let $f(x)$ be continuous and assumes only irrational values. Let $x_1 < x_2$. To prove that $f(x)$ is constant, we must prove that $f(x_1) = f(x_2)$.

If possible suppose $f(x_1) \neq f(x_2)$, say $f(x_1) < f(x_2)$, where $f(x_1$ and $f(x_2)$ are unequal irrationals. Then by Rational Density Theorem (8.4), there exists a rational element such that $f(x_1) < \mu < f(x_2)$.

Consider $[x_1, x_2]$. Then $f(x)$ is continuous in $[x_1, x_2]$ and $f(x_1) < \mu < f(x_2)$. Therefore, by Intermediate Value Theorem (3.9), there exists an element c in $[x_1, x_2]$ such that $f(c) = \mu$, which is rational This is a contradiction because $f(x)$ assumes only irrational values. Hence our supposition that $f(x_1) < f(x_2)$ must be wrong. Similarly we can prove that $f(x_1) > f(x_2)$ is wrong.

Thus $f(x_1) = f(x_2)$; i.e. the function is constant.

Exercise 3.15. If $f(x) = x\sin\frac{1}{x}$, $x \neq 0$, and
$f(x) = 0$, $x = 0$.
Then prove that $f(x)$ is continuous at 0.

Solution: It is given that
$f(x) = x\sin\frac{1}{x}$, $x \neq 0$, and
$f(x) = 0$, $x = 0$.
Therefore, $f(0) = 0$

Now

$$|f(x) - 0| = |f(x) - f(0)| = |f(x)| = |x\sin\tfrac{1}{x}| = |x||\sin\tfrac{1}{x}| \leq |x|$$

(because $|\sin\frac{1}{x}| \leq 1$)

We choose $\epsilon = \delta$ and $|x| < \delta$. Then

$$|f(x) - f(0)| < \epsilon, \text{ for } |x - 0| < \delta,$$

which proves that $f(x)$ is continues at 0.

Exercise 3.16. Suppose $f(x)$ is continuous in $[a, b] = I$ and assumes every value exactly once in I, then $f(x)$ is monotonic in I, and so is convergent.

Solution: Suppose $f(x)$ is continues in $[a, b] = I$. Let $a < x < b$. In order to prove that $f(x)$ is monotonic in I, we will prove that

$$f(a) < f(x) < f(b) \tag{85}$$

If possible suppose that result is false. Then negation of (86) is either $f(a) = f(x)$ which is not possible because by hypothesis $f(x)$ assumes every value exactly once.

Therefore $f(a) > f(x_1)$.

Consider $g(x) = f(x) - f(a)$, then $g(x)$ is continues in $[a, b] = I$ (because $f(x)$ is given to be continuous in $[a, b] = I$)

Now $g(x_1) = f(x_1) - f(a)$

< 0 (because $f(x_1) < f(a)$) and

$g(b) = f(b) - f(a)$

> 0 (because $f(b) > f(a)$)

Thus $g(x)$ is continues in $[a, b] = I$ and $g(x_1)$ and $g(b)$ differ in sign, therefore there must exist a point in I at which $g(x) = 0$ i.e. $f(x) = f(a)$ at that point which is not possible because the function $f(x)$ assumes every value exactly once in I.

Therefore, $f(a) \not> f(x_1)$ and so $f(a) < f(x_1)$

Also the negation of (86) is either $f(x_1) = f(b)$ which is impossible because $f(x)$ assumes every value exactly once in I.

or $f(x_1) > f(b)$

Consider $g(x) = f(x) - f(b)$, then $g(x)$ is continuous in I because so is $f(x)$.

Also $g(a) = f(a) - f(b) < 0$ (because $f(a) < f(b)$)

and $g(x_1) = f(x_1) - f(b) > 0$ (because $f(x_1) > f(b)$

Thus $g(x)$ is continuous in $[a, x_1]$ (because $g(x)$ is continuous in I) and $g(a)$ and $g(x_1)$ differ in sign.

therefore $g(x)$ must vanish for at least one x in $[a, x_1]$ i.e. $g(x)$ must vanish for at least one x in I.

or $f(x) = f(b)$ for at least one x in I which is clearly a contradiction because by hypothesis $f(x)$ assumes every value exactly once in I. Hence our supposition must be wrong.

Therefore, $f(x_1) \not> f(b)$

Thus if $a < x_1 < b$ then $f(a) < f(x_1) < f(b)$

Hence from (86) it follows that if $a < x_1 < x_2 < b$, then,

$$f(a) < f(x_1) < f(x_2) < f(b)$$

i.e if $x_1 < x_2$, then $f(x_1) < f(x_2)$ which proves that the function is monotonic.

Exercise 3.17. If $f(x) = x\sin\frac{1}{x}$, $x \neq 0$, and
$f(x) = 0$, $x = 0$.
Then prove that $f(x)$ is continuous at 0.

Solution: It is given that
$f(x) = x\sin\frac{1}{x}$, $x \neq 0$, and
$f(x) = 0$, $x = 0$.
Therefore, $f(0) = 0$

Now

$$|f(x) - 0| = |f(x) - f(0)| = |f(x)| = |x \sin\tfrac{1}{x}| = |x||\sin\tfrac{1}{x}| \leq |x|$$

(because $|\sin\tfrac{1}{x}| \leq 1$)

We choose $\epsilon = \delta$ and $|x| < \delta$. Then

$$|f(x) - f(0)| < \epsilon, \text{ for } |x - 0| < \delta,$$

which proves that $f(x)$ is continues at 0.

Exercise 3.18. Suppose $f(x)$ is continuous in $[a, b] = I$ and assumes every value exactly once in I, then $f(x)$ is monotonic in I, and so is convergent.

Solution: Suppose $f(x)$ is continues in $[a, b] = I$. Let $a < x < b$. In order to prove that $f(x)$ is monotonic in I, we will prove that

$$f(a) < f(x) < f(b) \tag{86}$$

If possible suppose that result is false. Then negation of (86) is either $f(a) = f(x)$ which is not possible because by hypothesis $f(x)$ assumes every value exactly once.

Therefore $f(a) > f(x_1)$.

Consider $g(x) = f(x) - f(a)$, then $g(x)$ is continues in $[a, b] = I$ (because $f(x)$ is given to be continuous in $[a, b] = I$)

Now $g(x_1) = f(x_1) - f(a)$

< 0 (because $f(x_1) < f(a)$) and

$g(b) = f(b) - f(a)$

> 0 (because $f(b) > f(a)$)

Thus $g(x)$ is continues in $[a, b] = I$ and $g(x_1)$ and $g(b)$ differ in sign, therefore there must exist a point in I at which $g(x) = 0$ i.e. $f(x) = f(a)$ at that point which is not possible because the function $f(x)$ assumes every value exactly once in I.

Therefore, $f(a) \ngtr f(x_1)$ and so $f(a) < f(x_1)$

Also the negation of (86) is either $f(x_1) = f(b)$ which is impossible because $f(x)$ assumes every value exactly once in I.

or $f(x_1) > f(b)$

Consider $g(x) = f(x) - f(b)$, then $g(x)$ is continuous in I because so is $f(x)$.

Also $g(a) = f(a) - f(b) < 0$ (because $f(a) < f(b)$)

and $g(x_1) = f(x_1) - f(b) > 0$ (because $f(x_1) > f(b)$

Thus $g(x)$ is continuous in $[a, x_1]$ (because $g(x)$ is continuous in I) and $g(a)$ and $g(x_1)$ differ in sign.

therefore $g(x)$ must vanish for at least one x in $[a, x_1]$ i.e. $g(x)$ must vanish for at least one x in I.

or $f(x) = f(b)$ for at least one x in I which is clearly a contradiction because by hypothesis $f(x)$ assumes every value exactly once in I. Hence our supposition must be wrong.

Therefore, $f(x_1) \ngtr f(b)$

Thus if $a < x_1 < b$ then $f(a) < f(x_1) < f(b)$

Hence from (86) it follows that if $a < x_1 < x_2 < b$, then,

$$f(a) < f(x_1) < f(x_2) < f(b)$$

i.e if $x_1 < x_2$, then $f(x_1) < f(x_2)$ which proves that the function is monotonic.

Remark 3.19. Had we supposed $a > x_1 > b$ then $f(a) > f(x_1) > f(b)$ and therefore in that case if $x_1 > x_2$, then we would get $f(x_1) > f(x_2)$.

Corollary 3.20. *Let* $f(x) = lim_{n\to\infty} \frac{log(2+x)-x^{2n} sinx}{1+x^{2n}}$ *in* $[0, \frac{\pi}{2}]$. *Prove that* $f(0)$ *and* $f(\frac{\pi}{2})$ *differ in sign. Also* $f(x)$ *does not vanish for any* x *between* 0 *and* $\frac{\pi}{2}$.

Proof. It is given that $f(x) = lim_{n\to\infty} \frac{log(2+x)-x^{2n} sinx}{1+x^{2n}}$ in $[0, \frac{\pi}{2}]$. Since $\frac{\pi}{2} > 1$, we break $[0, \frac{\pi}{2}]$ in three subintervals $0 \le x < 1$, $x = 1$, and $1 < x \le \frac{\pi}{2}$. Now if $0 \le x < 1$, then $x^{2n} \to 0$ as $n \to \infty$. Therefore $f(x) = log(2 + x)$

Thus

$$f(x) = log(2 + x); 0 \le x < 1 \tag{87}$$

Now at $x = 1$, $f(x) = \frac{log3 - sin1}{2}$. Thus

$$f(x) = \frac{log3 - sin1}{2} \tag{88}$$

Now for $1 < x \le \frac{\pi}{2}$.

$$f(x) = lim_{n\to\infty} \frac{log(2+x)-x^{2n} sinx}{1+x^{2n}}$$

$$f(x) = lim_{n\to\infty} \frac{\frac{log(2+x)}{x^{2n}} - Sinx}{\frac{1}{x^{2n}}+1}$$

$f(x) = sinx$ (because for $1 < x \le \frac{\pi}{2}$, $\frac{1}{x^{2n}} \to 0$ as $n \to 0$)

Thus

$$f(x) = -sinx; 1 < x \le \frac{\pi}{2} \tag{89}$$

Now from (87), $f(0) = log2 > 0$, and from (89) $f(\frac{\pi}{2}) = -sin(\frac{\pi}{2}) = -1 < 0$. Thus $f(0)$ and $f(\frac{\pi}{2})$ differ in sign.

Also from (87), for $0 \leq x < 1$, $f(x) = log(2+x) > log2 \neq 0$ (because $x > 0$, therefore $log(2+x) > log2 \neq 0$).

Now from (88), for $x = 1$, $f(x) = \frac{log3 - sin1}{2} \neq 0$ (because $log3 \neq sin1$), and from (89) for, $1 < x \leq \frac{\pi}{2}$, $f(x) = -sinx \neq 0$ (because $sinx = 0$ at $x = 0$ and $x = \pi$; and 0 and π do not lie in $1 < x \leq \frac{\pi}{2}$).

From above it follows that $f(x)$ does not vanish for any x in $[0, \frac{\pi}{2}]$. $\square$

Remark 3.21. In above result we have proved that $f(0)$ and $f\frac{\pi}{2}$ differ in sign and $f(x)$ does not vanish in $[0, \frac{\pi}{2}]$. The reason is that $f(x)$ is discontinuous at $x = 1$ as $f(1) = \frac{log3 - sin1}{2}$, and

for $h > 0$, $f(1-h) = log(2 + 1 - h) = log(3 - h)$, therefore,

$lim_{h \to 0} f(1 - h) = log3$, and

for $h > 0$, $f(1 + h) = -sin(1 + h)$, therefore,

$lim_{h \to 0} f(1 + h) = -sin1$

$lim_{h \to 0} f(1 - h) \neq lim_{h \to 0} f(1 + h)(\neq f(1))$.

Exercise 3.22. If $f(x) = x$, x is rational

$f(x) = -x$, x is irrational.

Then prove that $f(x)$ is continuous only at the point 0.

Solution: It is given that

$f(x) = x$, x is rational

$f(x) = -x$, x is irrational.

Let α be rational. Then $f(\alpha) = \alpha$.

Also for $h > 0$, $f(\alpha + h) = (\alpha + h)$ or $-(\alpha + h)$; according as h is rational or irrational.

Therefore, $\lim_{h \to 0} f(\alpha + h) = \alpha$ or $-\alpha$

$\alpha = -\alpha$ if and only if $\alpha = 0$.

Now $f(0) = 0$————(i)

For $h > 0$, $f(0 + h) = (0 + h)$ or $-(0 + h)$ according as h is rational or irrational.

Therefore, $\lim_{h \to 0} f(0 + h) = 0$ ————(ii)

Similarly $\lim_{h \to 0} f(0 - h) = 0$————-(iii)

From (i), (ii) and (iii), it follows that $f(x)$ is continues at $x = 0$ and if $\alpha \neq 0$, then $\lim_{h \to 0} f(\alpha + h)$ does not exist and similarly $\lim_{h \to 0} f(\alpha - h)$ does not exist which proves that $f(x)$ is discontinuous at every rational point other than 0.

Now let β be irrational, then $f(\beta) = -\beta$

Also for $h > 0$, $f(\beta + h) = (\beta + h)$ or $-(\beta + h)$ depending on h.

Therefore, $\lim_{h \to 0} f(\beta + h) = \beta$ depending on h.

Since for irrational β, $\beta \neq -\beta$, therefore, it follows that $\lim_{h \to 0} f(\beta + h)$ does not exist.

Similarly $\lim_{h \to 0} f(\beta - h)$ does not exist, which proves that $f(x)$ is discontinuous at every irrational point.

Thus $f(x)$ is discontinuous everywhere except at the point $x = 0$.

Theorem 3.23. Existence of nth root of positive numbers: *Let $k > 0$ and $n \geq 1$ an integer. Then there is one and only one c such that $c^n = k$.*

Proof. Let $k > 0$ and $n \geq 1$ an integer. For $n = 1, c = k$; and for $k = 1$, $c = 1$. Thus these are two trivial cases.

We now prove that there is one and only one such c. If possible suppose that there are two such c's say c_1, c_2; $c_1 > 0$, $c_2 > 0$, then $c_1^n = k$ and $c_2^n = k$

or, $c_1^n = c_2^n$, which implies that $c_1 = c_2$

Now let $k > 1$; and consider $f(x) = x^n - k$. Then $f(x)$ is continuous everywhere, therefore $f(x)$ is continuous in $[1, k]$. Now $f(1) = 1 - k < 0$, and $f(k) = k^n - k > 0$ (because $k > 1$, therefore, $k^n > k$).

Thus $f(1)$ and $f(k)$ differ in sign and $f(x)$ is continuous in $[1, k]$, therefore, there exists some c, $1 < c < k$ such that $f(c) = 0$, or $f(c) = c^n - k$; i.e. $c^n = k$.

For $0 < k < 1$, consider $f(x) = x^n - k$. Then f is continuous everywhere, therefore $f(x)$ is continuous in $[k, 1]$. Now $f(1) = 1 - k > 0$, and $f(k) = k^n - k$ (because $k < 1$, therefore $k^n < k$). Thus $f(x)$ is continuous in $[k, 1]$; $f(1)$ and $f(k)$ differ in sign, therefore there exists some c, $k < c < 1$ such that $f(c) = 0$; i.e. $c^n = k$. $\qquad\square$

Definition 3.24. Uniform continuity: A function $f(x)$ is said to be uniformly continuous in $[a, b] = I$ if given $\epsilon > 0$, we can find $\delta > 0$ depending only on ϵ such that $|f(x) - f(\alpha)| < \epsilon$ for $|x - \alpha| < \delta$, where α is any point in $[a, b] = I$.

Example:

Consider $f(x) = x^2$, $x \in [0, l]$. choose $\delta = \frac{\epsilon}{2l}$. Then δ depends only on ϵ. Let a be any point in $[0, l]$. Then

$$|f(x) - f(a)|$$

$$= x^2 - a^2$$

$$= |x + a||x - a|$$

$$< 2l|x - a|$$

$$< 2l\frac{\epsilon}{2l}$$

$$< \epsilon.$$

Therefore, $|f(x) - f(a)| < \epsilon$, for $|x - a| < \delta$, where a is any point in $[0, l]$. Now $\delta > 0$ depends only on ϵ, therefore, $f(x)$ is uniformly continuous.

Theorem 3.25. *(Hien's Theorem) Suppose $f(x)$ is continuous in $[a, b] = I$, then it is uniformly continuous in I.*

Proof. suppose $f(x)$ is continuous in $[a, b] = I$. Then given $\epsilon > 0$ we can divide I in finite number of of closed subintervals such that oscillation of f in any close subinterval, whose length does not exceed δ, is less than ϵ.

Now let $\delta = \mathrm{Max}((x_1 - x_0), (x_2 - x_1), ..., (x_n - x_{n-1}))$

where $a = x_0 < x_1 < x_2 < ... < x_{n-1} = b$ are points of partition of I.

Since δ is finite number of positive numbers, therefore $\delta > 0$ and δ depends upon ϵ

Let x' and x'' be any two points in I such that $|x' - x''| < \delta$.

Then there are two possibilities either x' and x'' lie in the same sub interval or in two successive subintervals.

If x' and x'' lie in the same subinterval such that $|x' - x''| < \delta$, then $|f(x') - f(x'')| < \epsilon$, so that result is proved in that case.

Now if x' and x'' lie in two successive close subintervals say $x' \in [x_{r-1}, x_r, \, x'' \in [x_r, x_{r+1}]$

then $|f(x') - f(x'')| = |f(x') - f(x_r) + f(x_r - f(x'')|$

$$\leq |f(x') - f(x_r)| + |f(x_r - f(x'')|$$

$< \epsilon + \epsilon$ (because $|f(x') - f(x_r)| < \epsilon$ and $|f(x_r - f(x'')| < \epsilon$)

Thus for any $x', x'' \in I$ such that $|x' - x''| < \delta$ where x' and x'' lie in two successive close subintervals, we have $|f(x') - f(x'')| < 2\epsilon$

This shows that $f(x)$ is uniformly continues in I. Hence the result follows. □

Exercise 3.26. Suppose $f(x)$ is continuous in $[a, b] = I$ and $f(x) = 0$ for all rational $x \in I$. Prove that $f(x) \equiv 0$.

Solution: It is given that $f(x)$ is continuous in $[a, b] = I$ and $f(x) = 0$ for all rational $x \in I$. We have to prove that $f(\alpha) = 0$ for all irrational α.

Let α be an irrational number, $\alpha \in I$, then there exists a sequence of rational $x_1, x_2, ..., x_n, ...$ in I such that

$$lim_{n\to\infty} x_n = \alpha \tag{90}$$

Since f is continuous in $[a, b] = I$, therefore from (90), we have

$$lim_{n\to\infty} f(x_n) = f(\alpha)$$

But $f(x_n) = 0$ [because $x_1, x_2, ..., x_n, ...$ are rational in I and $f(x) = 0$ for all rational $x \in I$]

Therefore, $f(\alpha) = 0$ for all irrational α.

This proves the result.

Theorem 3.27. *Suppose $f(x)$ is continues in $[a, b] = I$ and E is a set of points in I such that $x \in E$ if and only if $f(x) = 0$. Prove that E is closed.*

Proof. It is given that $f(x)$ is continues in $[a, b] = I$ and E is a set of points in I such that $x \in E$ if and only if $f(x) = 0$.

Let x_0 be a limit point of E.

Since $E \subseteq I$, therefore x_0 is a limit point of I, therefore there exists a sequence $x_1, x_2, ..., x_n, ...$ in E such that

$$lim_{n\to\infty} x_n = x_0 \tag{91}$$

Now f is continuous in $[a, b] = I$, therefore $f(x)$ is continuous in E (because $E \subseteq I$)

Therefore, by continuity of f, it follows from (91) that $lim_{n \to \infty} x_n = x_0$

or, $f(x_0) = 0$ (because $f(x_n) = 0$ as $x_n \in E$; for all n and $x \in E$ if and only if $f(x) = 0$).

Therefore, $x_0 \in E$; which proves that E is closed. $\square$

Exercise 3.28. $f(x) = \frac{1}{a}$, $x = \frac{p}{q}$ (in its lowest form)

$= 0$, x is irrational, $x \in (0, 1] = I$.

Prove that if f is discontinuous at every rational point and continuous at every irrational point in I.

Solution: It is given that

$f(x) = \frac{1}{a}$, $x = \frac{p}{q}$ (in its lowest form)

$=$, x is irrational, $x \in (0, 1] = I$.

We first prove that this function is discontinuous at every rational point.

Now $f(\frac{p}{q}) = \frac{1}{q}$

$f(\frac{p}{q+h}) = 0$, if h is irrational

$f(\frac{p}{q-h}) = 0$, if h is irrational.

Thus $f(\frac{p}{q \pm h}) = 0$, if h is irrational

and $f(\frac{p}{q \pm h}) \neq 0$, if h is rational.

Therefore, $lim_{h \to 0}(f(p/q \pm h))$ does not exist. So $f(x)$ is discontinuous at $\frac{p}{q}$ i.e. $f(x)$ is discontinuous at all rational points.

We now prove that $f(x)$ is continuous at at every irrational point.

Let α be any irrational number. Then $f(\alpha) = 0$.

To prove the result we must prove that given $\epsilon > 0$ there exist $\delta > 0$ such that $|f(x) - f(\alpha)| < \epsilon$ for $|x - \alpha| < \delta$. i.e. We must prove that

$$|f(x)| < \epsilon \tag{92}$$

for $|x - \alpha| < \delta$ [because $f(\alpha) = 0$].

Now if x is irrational, then $f(x) = 0$. Therefore, we must show that $0 < \epsilon$, for $|x - \alpha| < \delta$ which is always true.

Therefore, if x is irrational, then (92) is true and so the result is proved in this case.

Now let x be rational say $x = \frac{p}{q}$ in its lowest form, where $p \leq q$ as $x \in [0, 1]$.

Then to prove this result we must prove that

$$f(\frac{p}{q}) = \frac{1}{q} < \delta \tag{93}$$

Now let $\epsilon > 0$ be given. We think of all rational numbers $\frac{p}{q}$ in I. Let n be a positive integer and $n > \frac{1}{\epsilon}$. We fix n and then we think of all rational numbers $\frac{p}{q}$ whose denominator doe not exceed n, i.e. $q \leq n$. Therefore $p \leq n$. Such rational numbers will be finite in number.

Therefore it follows that one of these numbers say $\frac{p'}{q}$ is nearest to α.

Choose $\delta = |\frac{p'}{q} - \alpha|$, then $\delta > 0$, and the number of rationals in I which are still closer to α is infinite. Therefore, their denominator must exceed n. So if $\frac{p}{q}$ is in its lowest form such that $|\frac{p}{q} - \alpha| < \delta$ then $q > n$, i.e. $\frac{1}{q} < \frac{1}{n} < \epsilon$ for $|\frac{p}{q} - \alpha| < \delta$ which is (93).

Therefore, from above it follows that $f(x)$ is continuous at all irrational points.

Exercise 3.29. Consider the sum to n terms of the series and find the limit function and discuss it for continuity $\frac{x}{(x+1)(2x+1)} + \frac{x}{(2x+1)(3x+1)} + \frac{x}{(3x+1)(4x+1)} + \cdots$

Solution: The given series is $\frac{x}{(x+1)(2x+1)} + \frac{x}{(2x+1)(3x+1)} + \frac{x}{(3x+1)(4x+1)} + \ldots$

$$t_n(x) = \frac{x}{[(nx+1)][(n+1)x+1]}$$

$$= \frac{1}{nx+1} - \frac{1}{(n+1)x+1} \text{ (by partial fractions)}$$

Therefore, $t_1(x) = \frac{1}{x+1} - \frac{1}{2x+1}$

$$t_2(x) = \frac{1}{2x+1} - \frac{1}{3x+1}$$

$$t_3(x) = \frac{1}{3x+1} - \frac{1}{4x+1}$$

... ...

... ...

$$t_n(x) = \frac{1}{nx+1} - \frac{1}{(n+1)x+1}$$

Therefore, $S_n(x) = t_1(x) + t_2(x) + \ldots + t_n(x)$

$$= \frac{1}{x+1} - \frac{1}{(n+1)x+1}$$

Therefore, $lim_{n \to \infty} S_n(x) = \frac{1}{1+x}$

This function is continuous every where except at the point $x = -1$.

Exercise 3.30. Consider the sum to n terms of the series, find the limit function and discuss for continuity $\frac{x}{(x+1)} + \frac{x}{(x+1)(2x+1)} + \frac{x}{(2x+1)(3x+1)} + \ldots$

Solution: The given series is

$$\frac{x}{(x+1)(2x+1)} + \frac{x}{(2x+1)(3x+1)} + \frac{x}{(3x+1)(4x+1)} + \ldots$$

n^{th} term of series is given by

$$t_n x = \frac{x}{[(n-1)x+1][nx+1]}$$

or, $t_n x = \frac{1}{(n-1)x+1} - \frac{1}{nx+1}$ [by partial fraction]

Therefore, $t_1 x = 1 - \frac{1}{x+1}$

$t_2 x = \frac{1}{x+1} - \frac{1}{2x+1}$

$t_3 x = \frac{1}{2x+1} - \frac{1}{3x+1}$

... ...

... ...

$t_n x = \frac{1}{(n-1)x+1} - \frac{1}{nx+1}$

Therefore, $S_n(x) = t_1(x) + t_2(x) + ... t_n(x) = 1 - \frac{1}{(nx+1)}$.

Therefore, $lim_{n \to \infty} S_n(x) = 1$, which is continuous every where.

EXERCISES

(1) Show that $lim_{x\to 0}\dfrac{xe^{\frac{1}{x}}}{1+e^{\frac{1}{x}}} = 0$ and $lim_{x\to 0}\dfrac{e^{\frac{1}{x}}-1}{e^{\frac{1}{x}}+1}$ does not exist.

(2) Using the definition of limit, evaluate the following (if it exists):

     (a) $lim_{x\to 1}\dfrac{x^2+2x+5}{x^3+3x-8}$

     (b) $lim_{x\to 0}x^5\cos\dfrac{1}{x^2}$

     (c) $lim_{x\to 0}\dfrac{x}{1+e^{\frac{1}{x}}}$

     (d) $lim_{x\to 0}\dfrac{\sqrt{1+x}-\sqrt{1-x}}{x}$

(3) Prove the following:

     (a) $lim_{x\to 0}\dfrac{1-sec2x}{x^2} = 2$

     (b) $lim_{x\to 0}(1-x^2)^{\frac{1}{x}} = 0$

     (c) $lim_{x\to x}(1+2x)^{\frac{x+3}{3}} = e^6$

     (d) $lim_{x\to 1}sin\dfrac{\pi}{2}[x]$ does not exist.

(4) If $f(x) = lim_{x\to\infty}\dfrac{1}{1+nsin\pi x}$, then prove that

$$f(x) = 1, \text{ if } x \in \mathbb{Z} \text{ and } f(x) = 0, \text{ if } x \notin \mathbb{Z}$$

(5) If $f(x) = lim_{x\to\infty}(\frac{2}{\pi}tan^{-1}nx)$, then prove that

$$f(x) = 1 \text{ , if } x > 0$$
$$f(x) = 0, \text{ if } x = 0$$
$$f(x) = -1, \text{ if } x < 0$$

(6) Examine the continuity at $x = 1$;

$$f(x) = 2x, \text{ when } 0 \le x < 1$$
$$f(x) = 3, \text{ when } x = 1$$
$$f(x) = 4x, \text{ when } 1 < x \le 2$$

(7) Obtain the points of discontinuity of the function f, defined on $[0, 1]$ as follows:

$f(0) = 0$, $f(x) = \frac{1}{2} - x$, if $0 < x < \frac{1}{2}$,

$f(\frac{1}{2}) = \frac{1}{2}$, $f(x) = \frac{3}{2} - x$, if $\frac{1}{2} < x < 1$,

$f(1) = 1$

Also examine the kind of discontinuity.

(8) Test continuity of the following functions at the indicated points:

    (a) $f(x) = 0$, if $x^2 > 1$,

        $f(x) = 2$, if $x^2 < 1$,

        $f(x) = \frac{1}{2}$, if $x^2 = 1$,

        at $x = -1, 0$ and 1.

    (b) $f(x) = (1 + 2x)^{\frac{1}{x}}$, if $x \neq 0$,

        $f(x) = e^2$, if $x = 0$,

        at $x = 0$ and 1.

(9) Discuss the continuity of the following functions:

    (a) $f(x) = (x - 1)\sin\frac{1}{x-1}$, when $x \neq 1$; $f(x) = 0$, when $x = 1$
        over $\mathbb{R}$.

    (b) $f(x) = x^2 - x + 1$, when $2 \leq x \leq 1$; $f(x) = 5$, when $0 < x < 2$;
        $f(x) = x + 5$, when $-3 \leq x \leq 0$
        over $\mathbb{R}$

(10) Determine the discontinuities (if any), of the following functions:

    (a) $f(x) = x - [x]$

    (b) $f(x) = \frac{(x^2 - 3x + 2)}{(x^2 + x + 1)}$

    (c) $f(x) = e^{\frac{1}{x}} + \sin\frac{1}{x}$, if $x \neq 0$;
        $f(x) = 0$, if $x = 0$

(11) Give an example of function f defined and bounded over $[-1, 1]$ having

    (a) only one point of discontinuity

    (b) two points of discontinuity

    (c) three points of discontinuity

    (d) an infinite number of points of discontinuity

(12) Show that following functions are continuous but not uniformly continuous over indicated domains:

    (a) $f(x) = sin\frac{1}{x}$ over $[0, 1)$

    (b) $f(x) = \frac{1}{1-x}$ over $(0, 1)$

    (c) $f(x) = x^2$ over $\mathbb{R}$

(13) Show the following functions are uniformly continuous over indicated domains:

    (a) $f(x) = x^3 = 1$ over $(0, 2)$

    (b) $f(x) = \sqrt[n]{x}$ over $\mathbb{R}^+$, $n \in \mathbb{N}$.

Chapter4

DIFFERENTIABILITY

After having classified functions into two classes, continuous and discontinuous we shall further divide the class of continuous functions into two subclasses ; differentiable and non - differentiable.

The notion of differentiability is as basic as continuity but more useful than continuity. Both play a dominant role in entire calculus. continuity is a necessary condition for differentiability, but not a sufficient condition.

In this chapter, We shall study the derivative, its existence and applications. we shall be concerned mainly with real valued functions of a real variable i.e. the domain and ranges considered here will be sets of real numbers.

Definition 4.1. Suppose $f(x)$ is well defined at a. If $lim_{x \to a} \frac{f(x)-f(a)}{x-a}$ exists finitely, then $f(x)$ is said to be differentiable at the point $x = a$ and value of this limit is denoted by $f'(a)$.

Thus, $f'(a) = lim_{x \to a} \frac{f(x)-f(a)}{x-a}$.

It is clear from the definition that both right hand and left hand limits are involved i.e. for the right hand approach as $x \to a$ from the right the limit must exist finitely and limit must be equal to the limit as $x \to a$ from the left.

Example 4.2. Consider $f(x) = |x|$. Then $f(0) = 0$.

Suppose $a = 0$ here and $h > 0$, then for right hand approach $x = 0+h$ and for left hand approach $x = 0 - h$.

Therefore, $\frac{f(0+h)-f(0)}{0+h-0} = \frac{|0+h|}{h} - 0 = \frac{|h|}{h} = \frac{h}{h} = 1.$

Therefore $lim_{h\to 0}\frac{f(0+h)-f(0)}{h} = 1.$

i.e. right hand limit exists and is equal to 1.

Also for Left hand approach $x = 0 - h$

Therefore, $\frac{f(0-h)-f(0)}{0+h-0} = \frac{|0-h|}{-h} - 0 = \frac{|h|}{-h} = \frac{h}{-h} = -1.$

Therefore $lim_{h\to 0}\frac{f(0-h)-f(0)}{-h} = -1.$

i.e. left hand limit exists and is equal to -1.

Therefore $f(x)$ is not differentiable at $x = 0$.

i.e. $f'(0)$ does not exist.

On the other hand, we observe that $f(x)$ is continuous at $x = 0$. In fact that

$$lim_{h\to 0}f(0+h) = f(0) = lim_{h\to 0}f(0-h) = 0.$$

Thus, it follows from above that if a function is continuous at some point then it need not be differentiable there.

Theorem 4.3. *Suppose $f(x)$ is differentiable at $x = a$, then it is continuous at $x = a$.*

Proof. It is given that $f(x)$ is differentiable at $x = a$, therefore $f'(a)$ exists.

i.e.

$$lim_{h\to 0}\frac{f(x) - f(a)}{x - a} = f'(a) \tag{94}$$

Now, $f(x) - f(a) = \frac{f(x)-f(a)}{x-a}(x - a)$

Therefore, $lim_{x\to a}(f(x) - f(a)) = lim_{x\to a}\frac{f(x)-f(a)}{x-a}(x - a)$

$$= f'(a) \times 0$$

$$= 0 \text{ by } (94)$$

or, $lim_{x \to a} f(x) = f(a)$,

This shows that $f(x)$ is continuous at $x = a$. $\qquad\qquad\square$

Theorem 4.4. (Rolle's Theorem:) *Suppose $f(x)$ is continuous in $[a, b]$, differentiable in (a, b) and $f(a) = f(b)$. Then, there is at least one 'c' $a < b < c$ such that $f'(c) = 0$.*

Proof. It is given that $f(x)$ is continuous in $[a, b]$, differentiable in (a, b) and $f(a) = f(b)$. Since $f(x)$ is continuous in $[a, b]$, therefore it is bounded there, and hence

$$m \le f(x) \le M, \text{ for all } x \in [a, b]$$

If $m = M$, then $f(x)$ is constant.

Therefore, $f'(x) = 0$ for all $x \in [a, b]$ and the result is proved in that case.

Now let $m < M$, then $f(x)$ can not assume both m and M at the end points a and b because that leads to $m = M$.

Therefore, $f(c) = M$, for some c; $a < c < b$.

We will prove that $f'(c) = 0$

Let h be a small positive number. Then $f(c+h) \le f(c)$ and $f(c-h) \le f(c)$

or, $f(c + h) - f(c) \le 0$ and $f(c - h) - f(c) \le 0$. Therefore,

$$\frac{f(c + h) - f(c)}{h} \le 0 \qquad\qquad (95)$$

and

$$\frac{f(c-h)-f(c)}{-h} \geq 0 \tag{96}$$

Since $f(x)$ is given to be differentiable in (a,b) and $a < x < b$, therefore $f(x)$ is differentiable at c, i.e. $f'(c)$ exists.

Thus

$$f'(c) = lim_{h\to 0}\frac{f(c+h)-f(c)}{h} = lim_{h\to 0}\frac{f(c-h)-f(c)}{-h} \tag{97}$$

Letting $h \to 0$ in (95) and (96), We get

$$lim_{h\to 0}\frac{f(c+h)-f(c)}{h} \leq 0 \ lim_{h\to 0}\frac{f(c-h)-f(c)}{-h} \geq 0$$

Using (97), we get $f'(c) \leq 0$ and $f'(c) \geq 0$; $a < c < b$

Similarly if $f(c) = m$ for some c, $a < c < b$. We can prove that $f'(c) = 0$.

Hence the result. $\square$

Remark 4.5. The points c are the points of the relative maxima and minima of f in $[a,b]$.

Remark 4.6. Differentiability is not needed at end points of the interval. To see this consider

$$f(x) = \sqrt{1-x^2}, \ x \in [-1,1]$$

$$f'(x) = \frac{1}{2(1-x^2)^{-1/2}}2x$$

$$= \frac{-x}{\sqrt{1-x^2}} \text{ in } (-1,1)$$

Further $f(-1) = 0 = f(1)$. Thus $f(x)$ is continuous in $[-1,1]$

$f'(x)$ exists in $(-1,1)$ but is not defined at the end points -1 and 1 i.e. $f(x)$ is differentiable at end points -1 and 1 and $f(-1) = f(1)$ and the conclusion of Rolle's Theorem (4.4) holds for $x = 0$ as $f'(0) = 0$ and $-1 < 0 < 1$.

This proves that differentiability is not required at end points.

Remark 4.7. The hypothesis of Rolle's Theorem (4.4) is best possible.

To see this consider $f(x) = x,\ 0 \le x < 1;\ f(x) = 0,\ x = 1$.

Then $f'(x) = 1,\ 0 < x < 1$ i.e. $x \in (0, 1)$

Further $f(0) = 0 = f(1)$, but there does not exist any $c,\ 0 < c < 1$ such that $f'(c) = 0$ [because $f'(x) = 1,\ 0 < x < 1$]

So the conclusion of Rolle's Theorem (4.4) does not hold.

Thus $f(x)$ is differentiable in $(0, 1)$, $f(0) = f(1)$, but the conclusion of the Rolle's Theorem (4.4) does not hold. The reason is that $f(x)$ is discontinuous at $x = 1$ and hence in $[0, 1]$. In fact, $f(1) = 0$ and for $h > 0$, $f(1+h) = 1+h$, therefore $lim_{h \to 0} f(1+h) = 1$ and $f(1-h) = 1-h$ so that $lim_{h \to 0} f(1 - h) = 1$. Thus $lim_{h \to 0} f(1 \pm h) = 1 \ne f(1)$.

Remark 4.8. Consider the function $f(x) = |x|$. Then $f(x)$ is continuous in $[-1, 1]$. Further, $f(-1) = |-1| = 1$ and $f(1) = |1| = 1$ so that $f(-1) = f(1)$. But the Rolle's Theorem (4.4) is not applicable as $f(x)$ is not differentiable at $x = 0$ i.e. $f'(0)$ does not exist.

This shows that differentiability in open interval is necessary in Rolle's Theorem (4.4).

Remark 4.9. consider $f(x) = x,\ 0 \le x \le 1$. Then $f(x)$ is continuous in $[0, 1]$, differentiable in $(0, 1)$, but last condition of Rolle's Theorem (4.4) does not hold i.e. $f(0) \ne f(1)$ because $f(0) = 0,\ f(1) = 1$.

Thus the conclusion of the Rolle's Theorem (4.4) does not hold.

Exercise 4.10. Prove that between any two real roots of a real polynomial equation, lies a root of derived polynomial equation.

Solution: Let $P(x)$ be a polynomial with the real coefficient. Let $x_1, x_2,\ x_1 < x_2$ be the two real roots of $P(x)$. Therefore $P(x_1) = 0$ and $P(x_2) = 0$. So $P(x_1) = P(x_2)$

Now the polynomial $P(x)$ is continuous and differentiable every-where,

so, $P(x)$ is continuous in $[x_1, x_2]$ and differentiable in (x_1, x_2), also $P(x_1) = P(x_2)$.

Therefore, by Rolle's Theorem (4.4), it follows that there exists some c, $x_1 < c < x_2$ such that $P'(c) = 0$.

Thus, c is the root of the derived polynomial equation. Since c lies between x_1 and x_2.

Hence the result follows.

Theorem 4.11. *(**Lagrange's Mean Value Theorem***): Suppose $f(x)$ is continuous in $[a, b]$, differentiable in (a, b), Then there exists a 'c', $a < c < b$, such that*

$$f'(c) = \frac{f(b) - f(a)}{b - a}$$

Proof. It is given that $f(x)$ is continuous in $[a, b]$, differentiable in (a, b).

Consider $g(x) = f(x) - \mu x$, where μ is a constant to be chooser later. Now $g(x)$ is also continuous in $[a, \mu]$ and differentiable in (a, μ) {because μ being a constant is differentiable every where and hence continuous ev-erywhere, x being polynomial is also differentiable and hence continuous every where, therefore also in $[a, \mu]$. We want to apply Rolle's Theorem (4.4) to $g(x)$, first two conditions of Rolle's Theorem (4.4) are satisfied, last condition will be satisfied if $g(a) = g(b)$

i.e. if $f(a) - \mu a = f(b) - \mu b$

i.e. if $\mu b - \mu a = f(b) - f(a)$

i.e. if

$$\mu = \frac{f(b) - f(a)}{b - a} \tag{98}$$

With this value of μ, $g(a) = g(b)$. Thus $g(a)$ is continuous in $[a, b]$ and differentiable in (a, b) and $g(a) = g(b)$.

Therefore, by Rolle's Theorem (4.4), it follows that $g'(c) = 0$ for some c; $a < c < b$.

But $g'(c) = f'(c) - \mu$

i.e. $f'(c) - \mu = 0$ or $f'(c) = \mu$

using (98), it follows that $f'(c) = \frac{f(b)-f(a)}{b-a}$, $a < c < b$.

Geometrically this means, let there exists a point on the curve at which slope of the chord joining end points i.e. $(a, f(a))$ and $(b, f(b))$ will be equal to slope of the tangent at that point i.e. there will exist a point at which tangent to curve will be parallel to chord joining end points. $\square$

Theorem 4.12. *(**Cauchy Mean Value Theorem***): Suppose* $f(x)$ *and* $g(x)$ *are continuous in* $[a, b]$*; differentiable in* (a,b)*;* $g'(x) \neq 0$ *for* $a < x < b$*. Then* $\frac{f(b)-f(a)}{g(b)-g(a)} = \frac{f'(c)}{g'(c)}$*,* $a < c < b$*.*

Proof. It is given that $f(x)$ and $g(x)$ are continuous in $[a, b]$; differentiable in (a, b) and $g'(x) \neq 0$ for $a < x < b$. Consider $h(x) = f(x) - \lambda g(x)$, where λ is a constant to be chosen later.

Since $f(x)$ and $g(x)$ are continuous in $[a, b]$ and differentiable in (a, b), therefore $h(x)$ is continuous in $[a, b]$ and differentiable in (a, b). We want to apply Rolle's Theorem (4.4) to $h(x)$.

Now $h(a) = h(b)$ if $f(a) - \lambda g(a) = f(b) - \lambda g(b)$

i.e. if

$$\lambda = \frac{f(b) - f(a)}{g(b) - g(a)} \tag{99}$$

Therefore with this value of λ, $h(a) = h(b)$

Now $g(x)$ is continuous in $[a, b]$, differentiable in (a, b), therefore by Lagrange's Mean Value Theorem (4.11), it follows that there exists a c; $a < c < b$ such that $g(b) - g(a) = (b - a)g'(c)$.

Now $b - a \neq 0$ and $g'(x) \neq 0$ for $a < x < b$.

Therefore, $g(b) - g(a) \neq 0$ and so λ given by (99) is a well defined real number.

Using Rolle's Theorem (4.4) to the function $h(x)$ it follows that there exists a 'c' , $a < c < b$ such that $h'(c) = 0$ or $f'(c) - \lambda g'(c) = 0$

i.e. $\lambda = \dfrac{f'(c)}{g'(c)}$

$\Rightarrow \dfrac{f(b) - f(a)}{g(b) - g(a)} = \dfrac{f'(c)}{g'(c)}$, $a < c < b$ by using 99

Hence the result follows. $\qquad\qquad\qquad\qquad\qquad\qquad\qquad\qquad\square$

Remark 4.13. Taking $g(x) = x$ and $g'(c) = 1$ in the above result, we get

$\dfrac{f(b) - f(a)}{(b-a)} = f'(c)$ for $a < c < b$ which is Lagrange's Mean Value Theorem (4.11).

Theorem 4.14. (Taylor's Theorem). *suppose* $f(x)$, $f'(x)$, ..., $f^{n-1}(x)$ *are continuous in* $[a, b]$ *and* $f^n(x)$ *exists finitely in* (a, b). *Then*

$$f(b) = f(a) + (b - a)f'(a) + \frac{(b-a)^2}{2!}f''(a) + \ldots + \frac{(b-a)^{n-1}}{(n-1)!}f^{n-1}(a) + \frac{(b-a)^n}{n!}f^n(a), \ a < x < b.$$

Proof. It is given that suppose $f(x)$, $f'(x)$, ..., $f^{n-1}(x)$ are continuous in $[a, b]$ and $f^n(x)$ exists finitely in (a,b).

Let $f(b) = f(a) + (b - a)f'(a) + \frac{(b-a)^2}{2!}f''(a) + \ldots + \frac{(b-a)^{n-1}}{(n-1)!}f^{n-1}(a) + \frac{(b-a)^n}{n!}M.$

To prove the result, we will prove that $M = f^n(x)$, $a < x < b$

Now consider

$$g(t) = -f(b) + f(t) + (b-t)f'(t) + \frac{(b - t)^2}{2!}f''(t) + \ldots + \frac{(b - t)^{n-1}}{(n - 1)!}f^{n-1}(t) + \frac{(b - t)^n}{n!}f^n(t) \tag{100}$$

Now, $f(x)$, $f'(x)$, ..., $f^{n-1}(x)$ are continuous in $[a, b]$ and $(b - t)^n$ being a polynomial is also continuous in $[a, b]$.

Therefore, it follows from from (100), that $g(t)$ is also continuous in $[a, b]$.

Differentiate, we get,

$$g'(t) = f'(t) - f'(t) + (b - t)f''(t) - \frac{2(b-t)}{2!}f''(t) + \frac{2(b-t)^2}{2!}f'''(t) - \frac{3(b-t)^2}{3!}f'''(t) + ... - \frac{(n-1)(b-t)^{n-2}}{(n-1)!}f^{n-1}(t) + \frac{(b-t)^{n-1}}{n-1!}f^n(t) - \frac{n(b-t)^{n-1}}{n!}M$$

or $g'(t) = \frac{(b-t)^{n-1}}{n-1!}f^n(t) - \frac{n(b-t)^{n-1}}{n(n-1)!}f^n(t)M$.

So we have,

$$\Rightarrow g'(t) = \frac{(b - t)^{n-1}}{(n - 1)!}[f^n(t) - M] \tag{101}$$

Since $f^n(x)$ exists finitely in (a, b) and $(b-t)^{n-1}$ being a polynomial, it follows from above equation that $g'(t)$ exists in (a, b) i.e. $g(x)$ is differentiable in (a, b).

Also from (100), $g(a) = g(b) = 0$.

Thus $g(t)$ is continuous in $[a, b]$, differentiable in (a, b) and $g(a) = g(b)$.

Therefore, by Rolle's Theorem (4.4), it follows that there exists an x, $a < x < b$ such that $g'(x) = 0$.

i.e. $\frac{(b-x)^{n-1}}{(n-1)!}[f^x(t) - M] = 0$; using (101).

Now $\frac{(b-x)^{n-1}}{(n-1)!} \neq 0$ [because $b = x$ but $a < x < b$] Therefore it follows that $f^n(x) - M = 0$

or $M = f^n(x)$.

Hence substituting value of M, the results follows. $\qquad\square$

Remark 4.15.	(1) Suppose $f(x)$ and $g(x)$ are continuous in $[a, b]$, differentiable in (a, b) then applying Lagrange's Mean Value Theorem (4.11) to $f(x)$ and $g(x)$, it follows that there exists c_1, $a < c_1 < b$ such that

$$f(b) - f(a) = (b - a)f^{'}(c_1) \tag{102}$$

and there exists c_2, $a < c_2 < b$ such that

$$g(b) - g(a) = (b - a)g^{'}(c_2) \tag{103}$$

Dividing (102) and (103), we get

$$\frac{f(b) - f(a)}{g(b) - g(a)} = \frac{(b-a)f^{'}(c_1)}{(b-a)g^{'}(c_2)}; \; a < c_1 < b, \; a < c_2 < b$$

This show that Cauchy's Mean Value Theorem (4.12) does not follow directly from Lagrange's Mean Value Theorem (4.11).

(2) Taking $n = 2$ in Taylor's Theorem (4.14) in

$$f(b) = f(a) + (b - a)f^{'}(a) + \dots + \frac{(b-a)^{n-1}}{(n-1)!}f^{n-1}(a) + \frac{(b-a)^n}{n!}f^n(x)$$

we get

$$f(b) = f(a) + (b - a)f^{'}(c), \; a < c < b$$

or $f(b) - f(a) = (b - a)f^{'}(c)$, which is Lagrange's Mean Value Theorem (4.11).

(3) Taking $b = a + h$, $h > 0$ in the expansion of Taylor's Theorem (4.14), we get

$$f(a + h) = f(a) + (a + h - a)f^{(}a) + \dots + \frac{(a+h-a)^n}{n!}f^n(x), \; a < x < b$$

$$f(a + h) = f(a) + hf^{'}(a) + \frac{h^2}{2!}f^{''}(a) + \dots + \frac{h^n}{n!}f^n(a + \theta h), \; 0 < \theta < 1$$

Theorem 4.16. (Young's form of Taylor's Theorem): *If $f^n(x)$ exists finitely, then*

$$f(x + h) = f(x) + hf^{'}(x) + \frac{h^2}{2!}f^{''}(x) + \dots + \frac{h^n}{n!}(f^n(x) + \epsilon_h$$

where, $\epsilon_h \to 0$ as $h \to 0$.

Proof. It is given that $f^n(x)$ exists finitely.

Consider

$$\lambda(h) = f(x+h) - [f(x) + hf'(x) + \frac{h^2}{2!} + ... + \frac{h^n}{n!} f^{(}x)] \qquad (104)$$

and $\mu(h) = \frac{h^n}{n!}$.

Now for $h = 0$, $\frac{\lambda(h)}{\mu(h)}$ has $\frac{0}{0}$ form, $\frac{\lambda'(h)}{\mu'(h)}$ also has $\frac{0}{0}$ form for $h = 0$, $\frac{\lambda''(h)}{\mu''(h)}$ also has $\frac{0}{0}$ form for $h = 0$, ..., $\frac{\lambda^{n-1}(h)}{\mu^{n-1}(h)} \frac{0}{0}$ form for $h = 0$.

Now, $\lambda^n(h) = f^n(x+h) - f^n(x)$ all others do not exist, and $\mu^n(h) = \frac{n!}{n!} = 1$. (because for $f(x) = x^n$, $f(x^n) = n!$)

Now $lim_{h \to 0} \frac{\lambda^n(h)}{\mu^n(h)} = 0/1 = 0$.

Thus, $lim_{h \to 0} \frac{\lambda(h)}{\mu(h)} = 0$

Therefore $\frac{\lambda(h)}{\mu(h)} = \epsilon_h$, where $\epsilon_h \to 0$ as $h \to 0$

or $\lambda(h) = \epsilon_h \mu(h)$.

Now using this value of $\lambda(h)$ in (104), we get

$$\frac{\lambda(h)}{\mu(h)} = f(x+h) - [f(x) + hf' + \frac{h^2}{2!} f''(x) + ... + \frac{h^n}{n!} f^n(x)]$$

or $f(x+h) = f(x) + hf'(x) + \frac{h^2}{2!} f''(x) + ... + \frac{h^n}{n!} f^x$ (where $\epsilon_h \to 0$ as $h \to 0$) $\qquad \square$

Exercise 4.17. Suppose $f''(x)$ exists finitely. Then prove that
$$lim_{h \to 0} \frac{f(x+h) + f(x-h) - 2f(x)}{h^2} = f''(x)$$

Give an example to show that the converse is not true.

Solution: It is given that $f''(x)$ exists finitely.

Therefore, using Young's form of Taylor's Theorem (4.16), i.e. by taking $n = 2$, there, we get,

$$f(x + h) = f(x) + hf'(x) + \frac{h^2}{2!}(f''(x) + \epsilon_h), \text{ where } \epsilon_h \to 0 \text{ as } h \to 0$$

and

$$f(x - h) = f(x) - hf'(x) + \frac{h^2}{2!}(f''(x) + \epsilon_h'), \text{ where } \epsilon_h' \to 0 \text{ as } h \to 0$$

Therefore, $f(x + h) + f(x - h) = 2f(x) + \frac{h^2}{2!}[2f''(x) + \epsilon_h + \epsilon_h']$ or,

$$f(x + h) + f(x - h) - 2f(x) = h^2 f''(x) + \frac{h^2(\epsilon_h + \epsilon_h')}{2}$$

or, $\frac{f(x+h)+f(x-h)-2f(x)}{h^2} = f''(x) + \frac{\epsilon_h + \epsilon_h'}{2}$,

Letting $h \to 0$, we get

$$lim_{h \to 0} \frac{f(x+h)+f(x-h)-2f(x)}{h^2} = f''(x) + \frac{1}{2}lim_{h \to 0}(\epsilon_h) + \frac{1}{2}lim_{h \to 0}(\epsilon_h') = f''(x)$$

Therefore, $lim_{h \to 0} \frac{f(x+h)+f(x-h)-2f(x)}{h^2} = f''(x)$.

To see that converse is not true, consider the example

$$f(x) = x^2 sin(\tfrac{1}{x}), \ x \neq 0$$

$$f(x) = 0, \ x = 0$$

Then $f(0 + h) = h^2 sin(\tfrac{1}{h})$ and $f(0 - h) = -h^2 sin(\tfrac{1}{h})$.

Also $f(0) = 0$. So we have

$$f(0 + h) + f(0 - h) - 2f(0) = h^2 sin1\tfrac{1}{h} - h^2 sin(\tfrac{1}{h}) - 2.0 = 0.$$

Therefore, $lim_{h \to 0} \frac{f(0+h)+f(0-h)-2f(0)}{h^2} = 0$.

We now show that despite the fact that $lim_{h \to 0} \frac{f(0+h)+f(0-h)-2f(0)}{h^2}$ exists, $f''(x)$ does not exists.

Now, $f'(x) = 2x sin(1/x) - cos(1/x)$, $x \neq 0$. Also , $f(0) = 0$.

Therefore,

$$f'(0) = lim_{x \to 0} \frac{f(x)-f(0)}{x-0} = lim_{x \to 0} \frac{x^2 sin(1/x)}{x} = lim_{x \to 0} x\ sin(\tfrac{1}{x}) = 0.$$

Now $f''(0) = lim_{x \to 0} \frac{f'(x)-f'(0)}{x-0} = lim_{x \to 0} \frac{2x sin(\frac{1}{x}) - cos(\frac{1}{x}) - 0}{x} = $
$lim_{x \to 0}[2 sin(\tfrac{1}{x}) - (\tfrac{1}{x})\ cos(1x)]$.

Thus limit does not exist i.e. $f''(0)$ does not exist (because the term $lim_{x \to 0} sin(\tfrac{1}{x})$ on right side fails to exist. This can be seen as follows)

If we take a sequence of points $\{x_n\}_1^\infty$, where $x_n = \frac{1}{(2n+1)\frac{\pi}{2}}$, $n = 1, 2, 3, ...,$

then $lim_{n \to 0} x_n = 0$ and so

$$lim_{x \to 0} sin(\tfrac{1}{x}) = lim_{x \to 0}\ sin(2n + 1)\tfrac{\pi}{2} = \pm 1.$$

This shows that $lim_{x \to 0}\ sin(\tfrac{1}{x})$ does not exist. Hence $f''(0)$ does not exist.

Exercise 4.18. Prove that $\left|\frac{cos(\frac{\pi}{2})x}{log\frac{1}{x}}\right| < \frac{\pi x}{2}$, $x > 1$.

Solution: Consider $f(x) = cos(\frac{\pi x}{2})$ and $g(x) = log\frac{1}{x}$ in $[1, x]$ and differentiable in $(1, x)$. Also $g'(x) = -\frac{1}{x}$ (because $g(x) = log\frac{1}{x} = log1 - logx = -logx$).

Thus $g'(x) \neq 0$, in $[1, x]$.

Therefore, by Cauchy Mean Value Theorem (4.12), it follows that there exists a c , $1 < c < x$ such that

$$\frac{f(x)-f(1)}{g(x)-g(1)} = \frac{f'(c)}{g'(c)}$$

or, $\dfrac{\cos\frac{\pi}{2}x - \cos\frac{\pi}{2}}{\log\frac{1}{x} - \log 1} = \dfrac{-\frac{\pi}{2}\sin\frac{\pi c}{2}}{\frac{-1}{c}}$

or, $\dfrac{\cos\frac{\pi}{2}x}{\log\frac{1}{x}} = \dfrac{\pi c}{2}\sin\frac{\pi c}{2}$

Therefore, $\left|\dfrac{\cos(\frac{\pi}{2})x}{\log\frac{1}{x}}\right| = \left|\dfrac{\pi c}{2}\sin\frac{\pi c}{2}\right| = \dfrac{\pi c}{2}\left|\sin\frac{\pi c}{2}\right| \leq \dfrac{\pi}{2}.c.1$

So $\left|\dfrac{\cos(\frac{\pi}{2}x)}{\log\frac{1}{x}}\right| < \dfrac{\pi}{2}x$ (because $1 < c < x$).

Thus, $\left|\dfrac{\cos(\pi/2)x}{\log\frac{1}{x}}\right| < \dfrac{\pi x}{2}$, $x > 1$.

Theorem 4.19. (**Darboux's Theorem***): Suppose $f'(x)$ exists finitely in $[a, b]$,*

$$f'(a) < \mu < f'(b),$$

then there exists c, $a < c < b$ such that $f'(c) = \mu$.

Proof. It is given that $f'(x)$ exists finitely in $[a, b]$. Also

$$f'(a) < \mu < f'(b) \tag{105}$$

Consider $g(x) = f(x) - \mu x$, then,

$$g'(x) = f'(x) - \mu \tag{106}$$

Now by hypothesis $f'(x)$ exists finitely in $[a, b]$ and μ being a constant, it follows from (106) that $g'(x)$ exists finitely in $[a, b]$ i.e. $g(x)$ is differentiable in $[a, b]$ and hence continuous in $[a, b]$. Since $g(x)$ is continuous in $[a, b]$, it is bounded in $[a, b]$.

Now we know that a continuous function being bounded in a close interval assumes its bounds in the interval. Therefore, $g(x)$ will assume its greatest lower bound (say $= m$) in $[a, b]$.

We will show that $g(x)$ does not assume the value $'m'$ at the end points of interval.

Now $g'(x) = f'(x) - \mu$

i.e. $g'(a) = f'(a) - \mu < 0$ (because $f'(a) < \mu$)

Also, $g'(b) = f'(b) - \mu > 0$ (because $\mu < f'(b)$)

But $g'(a) = \lim_{h \to 0} \frac{g(a+h) - g(a)}{h} < 0$

Therefore, there must exist an $h > 0$, such that $\frac{g(a+h) - g(a)}{h} < 0$

or $g(a + h) - g(a) < 0$

or $g(a + h) < g(a)$

This clearly shows that $g(a) \neq m$

Thus $'m'$ can not be assumed at a.

Also $g'(b) = \lim_{h \to 0} \frac{g(b-h) - g(b)}{-h} > 0$

Therefore, there must exist an $h > 0$, such that $\frac{g(b-h) - g(b)}{-h} > 0$

or $\frac{g(b-h) - g(b)}{h} < 0$

or $g(b - h) - g(a) < 0$

or $g(b - h) < g(b)$

This shows that $g(b) \neq m$

i.e. $'m'$ is assumed at the points a and b.

Therefore, there exists a $'c'$, $a < c < b$ such that $g(c) = m$.

Therefore if, $h > 0$ is small real number then,

$$g(c + h) \geq g(c) \text{ and } g(c - h) \geq g(c)$$

or, $g(c+h) - g(c) \geq 0$ and $g(c-h) - g(c) \geq 0$

or, $\frac{g(c+h)-g(c)}{h} \geq 0$ and $\frac{g(c-h)-g(c)}{-h} \leq 0$.

Letting $h \to 0$, we get

$$lim_{h \to 0} \frac{g(c+h) - g(c)}{h} \geq 0 \tag{107}$$

and

$$lim_{h \to 0} \frac{g(c-h) - g(c)}{-h} \leq 0 \tag{108}$$

Now $g(x)$ is differentiable in $[a, b]$, therefore $g'(c)$ exists as $a < c < b$.

So from (107) and (108), we get $g'(c) \geq 0$ and $g'(c) \leq 0$ i.e. $g'(c) = 0$

But, $g'(c) = f'(c) - \mu$, therefore, $f'(c) - \mu = 0$

or, $f'(c) = \mu$, $a < c < b$.

Hence the result. $\qquad\qquad\qquad\qquad\qquad\qquad\qquad\qquad\qquad\qquad\quad$ $\square$

Corollary 4.20. *Let λ_1 and λ_2 be two positive real numbers. Then prove that $\lambda_1 f'(x_1) + \lambda_2 f'(x_2) = (\lambda_1 + \lambda_2) f'(x_3)$; where x_3 lies between x_1 and x_2.*

Proof. Suppose $f'(x_1) < f'(x_2)$

Then we claim that

$$f'(x_1) < \frac{\lambda_1 f'(x_1) + \lambda_2 f'(x_2)}{\lambda_1 + \lambda_2} < f'(x_2) \tag{109}$$

Now (109) is true if

$f' x_1 < \frac{\lambda_1 f'(x_1) + \lambda_2 f'(x_2)}{\lambda_1 + \lambda_2}$

i.e. if $\lambda_1 f'(x_1) + \lambda_2 f'(x_1) < \lambda_1 f'(x_1) + \lambda_2 f'(x_2)$

i.e. if $\lambda_2 f'(x_1) < \lambda_2 f'(x_2)$

i.e. if $f'(x_1) < f'(x_2)$, which is true.

Also (109) is true if

$$\frac{\lambda_1 f'(x_1) + \lambda_2 f'(x_2)}{\lambda_1 + \lambda_2} < f'(x_2)$$

i.e. if $\lambda_1 f'(x_1) + \lambda_2 f'(x_2) < \lambda_1 f'(x_2) + \lambda_2 f'(x_2)$

i.e. If $\lambda_1 f'(x_1) < \lambda_1 f'(x_2)$

i.e. if $f'(x_1) < f'(x_2)$ which is true.

Now let $\mu = \frac{\lambda_1 f'(x_1) + \lambda_2 f'(x_2)}{\lambda_1 + \lambda_2}$

Therefore, from (109), we get

$$f'(x_1) < \mu < f'(x_2)$$

Therefore by Darboux's Theorem (4.19), it follows that there exists an x_3 between x_1 and x_2 such that $f'(x_3) = \mu$

or, $\frac{\lambda_1 f'(x_1) + \lambda_2 f'(x_2)}{\lambda_1 + \lambda_2} = f'(x_3)$

i.e. $\lambda_1 f'(x_1) + \lambda_2 f'(x_2) = (\lambda_1 + \lambda_2) f'(x_3)$, where x_3 lies between x_1 and x_2. $\qquad\square$

Exercise 4.21. Suppose $c_0, c_1, c_2, ..., c_n$ are real numbers such that

$c_0 + \frac{c_1}{2} + \frac{c_2}{3} + ... + \frac{c_n}{n+1} = 0$, then prove that

$c_0 + c_1 x + c_2 x^2 + ... + c_n x^n = 0$ has at least one root between 0 and 1.

Solution: It is given $c_0, c_1, c_2, ..., c_n$ are real numbers such that

$c_0 + \frac{c_1}{2} + \frac{c_2}{3} + ... + \frac{c_n}{n+1} = 0.$

Consider the polynomial

$$P(x) = c_0 x + \tfrac{c_1}{2} x^2 + \tfrac{c_2}{3} x^3 + \ldots + \tfrac{c_n}{n+1} x^{n+1}.$$

Then $P(x)$ is differentiable and hence continuous everywhere. Thus $P(x)$ is continuous in $[0, 1]$ and differentiable in $(0, 1)$.

Also $P(0) = 0$ and $P(1) = c_0 + \tfrac{c_1}{2} + \tfrac{c_2}{3} + \ldots + \tfrac{c_n}{n+1} = 0$. (By hypothesis)

i.e. $P(0) = P(1)$

Therefore, applying Rolle's Theorem (4.4) to $P(x)$, it follows that there exists a k, $0 < k < 1$ such that $P'(k) = 0$

or, $c_0 + c_1(k) + c_2 k^2 + \ldots + c_n k^n = 0$

This shows that $x = k$, $0 < k < 1$ is a root of the equation $c_0 + c_1(x) + c_2 x^2 + \ldots + c_n x^n = 0$

Hence the result follows.

Exercise 4.22. If $f(x) = x$, x rational

$f(x) = sinx$, x irrational

Then prove that $f'(0)$ exists and is equal to 1.

Solution: It is given that $f(x) = x$, x rational

$f(x) = sinx$, x irrational

Now $f'(0) = lim_{x \to 0} \frac{f(x) - f(0)}{x - 0}$

$f'(0) = lim_{x \to 0} \frac{f(x)}{x}$ (because 0 is rational , therefore $f(0) = 0$).

Therefore, $f'(0)$ will exist if $lim_{x \to 0} \frac{f(x)}{x}$ exists.

If x is rational, then $f(x) = x$

Therefore, $lim_{x \to 0} \frac{f(x)}{x} = lim_{x \to 0} \frac{x}{x} = 1$.

Now if x is irrational, then $f(x) = sinx$

Therefore, $lim_{x \to 0} \frac{f(x)}{x} = lim_{x \to 0} \frac{sinx}{x} = 1$

Therefore, if $x \to 0$ through rational values or irrational values, then $lim_{x \to 0} \frac{f(x)}{x} = 1$.

Since $f'(0) = lim_{x \to 0} \frac{f(x)}{x}$

Therefore $f'(x)$ exists and is equal to 1.

Exercise 4.23. Suppose $f(x)$ is continuous in $[0, 1] = I$ and $f(I) = I$. Then prove that $f(x)$ has at least one fixed point.

Solution: It is given that $f(x)$ is continuous in $[0, 1] = I$ and $f(I) = I$. Therefore $0 \leq f(x) \leq 1$, $x \in I$.

If $f(0) = 0$, then 0 is the fixed point of f and if $f(1) = 1$, then 1 is fixed point of f. So that there is nothing to prove in this case.

Therefore, Let $0 < f(x) < 1$, in I.

Consider $g(x) = f(x) - x$

Then $g(x)$ is also continuous in $[0, 1]$ (because $f(x)$ is continuous in $[0, 1]$)

Now $g(1) = f(1) - 1 < 0$ and $g(0) = f(0) - 0 > 0$ (because $0 < f(x) < 1$

Therefore $g(x)$ is continuous in $[0, 1]$, $g(0)$ and $g(1)$ differ in sign. So there exists an x, $0 < x < 1$ such that $g(x) = 0$

i.e. $f(x) - x = 0$

i.e. $f(x) = x$.

Thus there exists an x, $0 < x < 1$ such that $f(x) = x$.

This shows that $f(x)$ has at least one fixed point.

Exercise 4.24. If $f(x) = e^{-\frac{1}{x^2}}, x \neq 0$

$f(x) = 0, \ x = 0$.

Prove that all the derivatives of $f(x)$ vanish at the origin and also show that Maclaurin's series of $f(x)$ converges to $f(x)$ at 0.

Solution: It is given that

$$f(x) = e^{-\frac{1}{x^2}}, x \neq 0$$

$$f(x) = 0, \ x = 0.$$

Now $f(0) = 0$

Also, $f'(0) = \lim_{x \to 0} \frac{f(x) - f(0)}{x - 0}$

$$= \lim_{x \to 0} \frac{e^{-\frac{1}{x^2}}}{x}$$

$$= \lim_{x \to 0} \frac{1}{x} \cdot \frac{1}{e^{\frac{1}{x^2}}}$$

$$= \lim_{x \to 0} \frac{1}{x} \left[\frac{1}{1 + \frac{1}{x^2} + \frac{1}{2!}(\frac{1}{x^2})^2 + \frac{1}{3!}(\frac{1}{x^2})^3} + \ldots \right]$$

$$= \lim_{x \to 0} \left[\frac{1}{x + \frac{1}{x} + \frac{1}{2!}(x^3) + \frac{1}{3!}(x^3) + \ldots} \right] = 0$$

Therefore, $f'(0) = 0$

Now, $f''(0) = \lim_{x \to 0} \frac{f'(x) - f'(0)}{x - 0}$

or $f''(0) = \lim_{x \to 0} \frac{\frac{1}{x^3} e^{\frac{-1}{x^2}} - 0}{x}$ (because $f'(0) = 0$)

or $f''(0) = \lim_{x \to 0} \frac{1}{x^4} \cdot \frac{1}{e^{\frac{1}{x^2}}}$

$$= lim_{x \to 0} \frac{1}{x^4} \left[\frac{1}{1 + \frac{1}{x^2} + \frac{1}{2!}(\frac{1}{x^2})^2 + \frac{1}{3!}(\frac{1}{x^2})^3} + \ldots \right]$$

$$= lim_{x \to 0} \left[\frac{1}{x^4 + \frac{1}{x^2} + \frac{1}{2!} + \frac{1}{3!}\frac{1}{x^2} + \ldots} \right] = 0$$

Therefore, $f''(0) = 0$.

Similarly $f'''(0) = 0$, $f^{iv}(0) = 0$, ...

Now Maclaurin's expansion of $f(x)$ is given by

$$f(x) = f(0) + x f'(0) + \frac{x^2}{2!} f''(0) + \ldots$$

$$= 0 + x.0 + \frac{x^2}{2!}.0 + \frac{x^3}{3!}.0 + 0 + \ldots$$

$$= 0$$

This shows that Maclaurin's series of $f(x)$ converges to $f(x)$ at $x = 0$.

Exercise 4.25. If $0 \leq h < 1$, then prove that

$$h < log\frac{1}{1-h} < \frac{h}{1-h}.$$

Solution: Let $f(h) = log(1 - h)$.

Then, $f(0) = log 1 = 0$

Also $f'(h) = \frac{1}{1-h}$, (here we note that $f'(h)$ exists because by hypothesis $0 \leq h < 1$)

Applying Lagrange's Mean Value Theorem to $f(h)$, and using the fact that $f(a + h) = f(a) + h f'(0 + \theta h), 0 < \theta < 1$, we get that

$$f(h) = f(0) + h f'(\vartheta h), \ 0 < \theta < 1$$

or $f(h) = 0 + h f'(\theta h)$

or $log(1 - h) = h f'(\theta h)$

or

$$log(1 - h) = \frac{-h}{1 - \theta h}, 0 < \theta < 1 \tag{110}$$

Now, since $h > 0$, $0 < \theta < 1$, therefore $\theta h > 0$, and so

$0 < 1 - \theta h < 1$

or $\frac{1}{1-\theta h} > 1$

or $\frac{-h}{1-\theta h} < -h$

Using this in (110) we get

$log(1 - h) < -h$

or $-log(1 - h) > h$

or $log\frac{1}{1-h} > h$

or

$$h < log\frac{1}{1 - h} \tag{111}$$

Now, $1 - \theta h > 1 - h$ if $-\theta h > -h$, i.e. if $\theta < 1$ which is true.

Therefore, $1 - \theta h > 1 - h$, which implies that

$\frac{1}{1-\theta h} < \frac{1}{1-h}$

or $\frac{-h}{1-\theta h} > \frac{-h}{1-h}$

Using this in (110), we get

$log(1 - h) > \frac{-h}{1-h}$

or $-log(1 - h) < \frac{h}{1-h}(-log(1 - h) = log(1 - h)^{-1} = log\frac{1}{1-h}$

or

$$log\frac{1}{1 - h} < \frac{h}{1 - h}. \tag{112}$$

Combining (111) and (112), it follows that

$$h < log\frac{1}{1-h} < \frac{h}{1-h}$$

Exercise 4.26. Suppose $f(x), g(x)$ and $h(x)$ satisfy the conditions of Lagrange's Mean Value Theorem in $[a, b] = I$, then prove that there exists a 'c', $a < c < b$ such that $\begin{vmatrix} f'(c) & g'(c) & h'(c) \\ f(a) & g(a) & h(a) \\ f(b) & g(b) & h(b) \end{vmatrix} = 0.$

Solution: It is given that $f(x), g(x)$ and $h(x)$ satisfy the conditions of Lagrange's mean value theorem i.e. $f(x), g(x)$ and $h(x)$ are continuous in $[a, b]$ and differentiable in (a, b).

$$\text{Now consider } F(x) = \begin{vmatrix} f(x) & g(x) & h(x) \\ f(a) & g(a) & h(a) \\ f(b) & g(b) & h(b) \end{vmatrix}$$

Then $F(x) = f(x)\{g(a)h(b) - g(b)h(a)\} - g(x)\{f(a)h(b) - h(a)f(b)\} + h(x)\{f(a)g(b) - g(a)f(b)\}$

Thus $F(x)$ is linear combination of the functions $f(x), g(x)$ and $h(x)$. Therefore $F(x)$ is continuous in $[a, b]$ and differentiable in (a, b).

$$\text{Now } F(a) = \begin{vmatrix} f(a) & g(a) & h(a) \\ f(a) & g(a) & h(a) \\ f(b) & g(b) & h(b) \end{vmatrix} = 0 \text{ (because two rows are identi-}$$

cal)

$$\text{and } F(b) = \begin{vmatrix} f(b) & g(b) & h(b) \\ f(a) & g(a) & h(a) \\ f(b) & g(b) & h(b) \end{vmatrix} = 0 \text{ (because two rows are identi-}$$

cal)

Therefore, $F(a) = F(b)$

Thus $F(x)$ is continuous in $[a, b]$, differentiable in (a, b) and $F(a) = F(b)$

Therefor, by Rolle's Theorem (4.4), there exists 'c', $a < c < b$ such that $F'(c) = 0$.

$$\text{But } F'(x) = \begin{vmatrix} f'(x) & g'(x) & h'(x) \\ f(a) & g(a) & h(a) \\ f(b) & g(b) & h(b) \end{vmatrix} + \begin{vmatrix} f(x) & g(x) & h(x) \\ 0 & 0 & 0 \\ f(b) & g(b) & h(b) \end{vmatrix}$$

$$+ \begin{vmatrix} f(x) & g(x) & h(x) \\ f(a) & g(a) & h(a) \\ 0 & 0 & 0 \end{vmatrix}$$

$$\text{Therefore, } F'(x) = \begin{vmatrix} f'(x) & g'(x) & h'(x) \\ f(a) & g(a) & h(a) \\ f(b) & g(b) & h(b) \end{vmatrix}.$$

$$\text{So } F'(c) = \begin{vmatrix} f'(c) & g'(c) & h'(c) \\ f(a) & g(a) & h(a) \\ f(b) & g(b) & h(b) \end{vmatrix} \text{ for } a < c < b$$

But $F'(c) = 0$ for $a < c < b$

Therefore from above it follows that

$$\begin{vmatrix} f'(c) & g'(c) & h'(c) \\ f(a) & g(a) & h(a) \\ f(b) & g(b) & h(b) \end{vmatrix} = 0$$

Exercise 4.27. Show that there is no function $f(x)$ such that $f'(x)$ exists for $x \geq 0$, $f'(0) = 0$ and $f'(x) \geq h$ for $x > 0$; $h > 0$.

Solution: If possible suppose a function $f(x)$ exists for $x \geq 0$

$f'(0) = 0$ and $f(x) \geq h$ for $x > 0$; $h > 0$

Since $f(x)$ satisfies the conditions of Lagrange's Mean Value Theorem. Therefore, applying Lagrange's Mean Value Theorem (4.13), we get

$$f(x) = f(0) + (x - 0)f'(\theta x) \, , \, 0 < \theta < 1$$

This gives $\qquad\qquad \dfrac{f(x) - f(0)}{x - 0} = f'(\theta x) \qquad\qquad\qquad (113)$

Since $\theta > 0$ and $x > 0$, so $\theta x > 0$ and hence

$$f'(\theta h) \geq h > 0 \tag{114}$$

(because by our supposition $f'(x) \geq h$ for $x > 0$)

Since $f'(x)$ exists for $x \geq 0$ and $f'(0) = 0$.

Therefore, letting $x \to 0$ in (113), we get

$$f'(0) = lim_{x \to 0} \frac{f(x)-f(0)}{x-0} = f'(\theta x) \geq h > 0 \text{ by using (114)}$$

i.e. $0 = f'(\theta x) \geq h > 0 \Rightarrow 0 > 0$, which is absurd.

Hence no such function $f(x)$ can exist such that $f'(x)$ exists for $x \geq 0$, $f'(0)$ and $f'(x) \geq h$ for $x > 0$.

Theorem 4.28. *suppose $f''(x)$ exists in $[a,b] = I$ and $\frac{f(c)-f(a)}{c-a} = \frac{f(b)-f(c)}{b-c}$, for some c, $a < c < b$, then prove that there exists a c' between a and b such that $f''(c') = 0$.*

Proof. It is given that $f''(x)$ exists in $[a,b] = I$ and $\frac{f(c)-f(a)}{c-a} = \frac{f(b)-f(c)}{b-c}$.

Therefore applying Taylor's Theorem (4.14)to $f(x)$, we get

$$f(a) = f(c) + (a-c)f'(c) + \frac{(a-c)^2}{2!}f''(c_1) \tag{115}$$

where $a < c_1 < b$

and

$$f(b) = f(c) + (b-c)f'(c) + \frac{(b-c)^2}{2!}f''(c_2) \tag{116}$$

where $a < c_2 < b$

From (115) and (116) we get

$$\frac{f(a) - f(c)}{a - c} = f'(c) + \frac{(a-c)}{2!}f''(c_1) \tag{117}$$

and

$$\frac{f(b) - f(c)}{b - c} = f'(c) + \frac{(b - c)}{2!} f''(c_2) \tag{118}$$

Now $\frac{f(a) - f(c)}{a - c} = \frac{f(b) - f(c)}{b - c}$ and therefore, from (117) and (118), we have

$$f'(c) + \frac{(a-c)}{2!} f''(c_1) = f'(c) + \frac{(b-c)}{2!} f''(c_2)$$

or $\frac{(a-c)}{2!} f''(c_1) = \frac{(b-c)}{2!} f''(c_2)$

or $\frac{(b-c)}{2!} f''(c_2) + \frac{(c-a)}{2!} f''(c_1) = 0.$

or $\lambda_1 f''(c_2) + \lambda_2 f''(c_1) = 0$

where $\lambda_1 = \frac{(b-c)}{2!} > 0$ and $\lambda_2 = \frac{(c-a)}{2!} > 0$

Therefore applying Corollary of Darboux's Theorem (4.20), it follows that there exists a c', $c_1 < c' < c_2$ such that

$$\lambda_1 f''(c_2) + \lambda_2 f''(c_1) = (\lambda_1 + \lambda_2) f''(c'), \; c_1 < c' < c_2$$

i.e. $((\lambda_1 + \lambda_2) f''(c')) = 0999$ or $(\frac{b-c}{2!} + \frac{c-a}{2!}) f''(c') = 0$

or $(\frac{b-a}{2!}) f''(c') = 0$

Now $\frac{b-a}{2} > 0$ implies that

$f''(c') = 0$, where $c_1 < c' < c_2$ because $a < c_1 < b$ and $a < c_2 < b$

Therefore, $f''(c) = 0$, $a < c' < b$ and the result follows. $\qquad\square$

Exercise 4.29. Suppose $f'''(x)$ exists in $[1, -1]$, $f(-1) = 0$, $f(0) = 0$, $f(1) = 1$, $f'(0) = 0$.

Then prove that there exists a point x, $-1 < x < -1$ such that $f'''(x) = 3$.

Solution: It is given that $f'''(x)$ exists in $[1, -1]$, $f(-1) = 0$, $f(0) =$

0, $f(1) = 1$, $f'(0) = 0$.

Applying Taylor's Theorem (4.14) to $f(x)$, we get

$$f(1) = f(0) + (1 - 0)f'(0) + \frac{(1 - 0)^2}{2!}f''(0) + \frac{(1 - 0)^3}{3!}f'''(c_1) \quad (119)$$

for some c_1, $0 < c_1 < 1$

and

$$f(-1) = f(0) + (-1 - 0)f'(0) + \frac{(-1 - 0)^2}{2!}f''(0) + \frac{(-1 - 0)^3}{3!}f'''(c_2) \quad (120)$$

for some c_2, $-1 < c_2 < 0$

From (119) and (120), we get

$$1 = 0 + (1 - 0)0 + \frac{(1-0)^2}{2!} + \frac{(1-0)^3}{3!}f'''(c_1), \ 0 < c_1 < 1$$

$$0 = 0 + (-1 - 0)0 + \frac{(-1-0)^2}{2!} + \frac{(-1-0)^3}{3!}f'''(c_2), \ -1 < c_2 < 0$$

(Because by hypothesis $f(-1) = 0$, $f(0) = 0$, $f(1) = 1$, $f'(0) = 0$)

or for $0 < c_1 < 1$

$$1 = \frac{1}{2}f''(0) + \frac{(1 - 0)^3}{3!}f'''(c_1), \quad (121)$$

and for $-1 < c_2 < 0$

$$0 = \frac{1}{2}f''(0) + \frac{(-1 - 0)^3}{3!}f'''(c_2) \quad (122)$$

Subtracting (121) from (122), we get

$$1 = \tfrac{1}{3!}(f'''(c_1 + f'''(c_2)), \ 0 < c_1 < 1, -1 < c_2 < 0$$

or $1 = \tfrac{1}{6}\{f'''(c_1) + f'''(c_2)\}$, $0 < c_1 < 1, -1 < c_2 < 0$

or

$$1.f'''(c_1) + 1.f'''(c_2) = 6 \quad (123)$$

$$0 < c_1 < 1, -1 < c_2 < 0$$

Applying Darboux's Theorem (4.19) to $f'''(x)$, it follows that there exists a c_3 between c_1 and c_2 such that

$$1.f'''(c_1) + 1.f'''(c_2) = (1+1)f'''(c_3)$$

or $6 = 2f'''(c_3)$ (using (123)), which implies that

$$f'''(c_3) = 3,$$

where c_3 lies between c_1 and c_2 and hence between -1 and 1.

Hence the result.

Exercise 4.30. Suppose $f'(x)$ exists finitely for $0 < x \leq 1$, $|f'(x)| < 1$, for $0 < x \leq 1$

Let $a_n = f(\frac{1}{n})$, $n = 1, 2, 3, \dots$

Prove that $\lim_{n \to \infty} a_n$ exists.

Solution: It is given that $f'(x)$ exists finitely for $0 < x \leq 1$, $|f'(x)| < 1$ for $0 < x \leq 1$ and $a_n = f(\frac{1}{n})$, $n = 1, 2, 3, \dots$

We have to show that $\lim_{n \to \infty} a_n$ exists.

To prove this, it suffices to prove that $\{a_n\}_1^\infty$ is a Cauchy sequence. (because $\{a_n\}_1^\infty$ is convergent if and only if it is a cauchy sequence.)

Now $|a_n - a_m| = |f(\frac{1}{n}) - f(\frac{1}{m})|$

Now $f(x)$ satisfies the conditions of Lagrange's Mean Value Theorem in close interval with end points $\frac{1}{n}$ and $\frac{1}{m}$. Therefore by Lagrange's Mean Value Theorem (4.13), it follows that

$$\frac{f(\frac{1}{n}) - f(\frac{1}{m})}{\frac{1}{n} - \frac{1}{m}} = f'(c), \text{ for some } c, \ \frac{1}{n} < c < \frac{1}{m}$$

or $f(\frac{1}{n}) - f(\frac{1}{m}) = (\frac{1}{n} - \frac{1}{m})f'(c), \ \frac{1}{n} < c < \frac{1}{m}$

$$|f(\frac{1}{n}) - f(\frac{1}{m})| \leq |\frac{1}{n} - \frac{1}{m}| \tag{124}$$

(because by hypothesis $|f'(x)| < 1, 0 < x \leq 1$)

(124) is also true for large m and n.

Also from (124), we get

$$|a_n - a_m| \leq |\frac{1}{n} - \frac{1}{m}| \tag{125}$$

(because $f(\frac{1}{n}) = a_n, n = 1, 2, ...$)

Now $lim\frac{1}{n} = 0$ implies that $\{\frac{1}{n}\}_1^\infty$ is convergent and hence a Cauchy sequence.

Therefore, given $\epsilon > 0$, we can find a positive integer N such that

$|\frac{1}{n} - \frac{1}{m}| < \epsilon$, for all $n, m \geq N$

Using this in (125), we get

$|a_n - a_m| < \epsilon$ for all $n, m \geq N$.

This shows that $\{a_n\}_1^\infty$ is a Cauchy sequence and hence convergent.

Therefore, $lim_{n \to \infty} a_n \, exists$

or $lim_{n \to \infty} a_n = lim_{n \to \infty} f(\frac{1}{n})$ exists.

Exercise 4.31. Suppose $g(x)$ is such that $g'(x)$ exists in a closed interval I and $|g'(x)| \leq m$ in I. Prove that
$f(x) = x + \epsilon g(x)$ is one - one, if $\epsilon > 0$ is choosen properly.

Solution: It is given that $g'(x)$ exists in I and $|g'(x)| \leq m$ in I.
Therefore $g(x)$ is differentiable and continuous in I.
Let $x_1, x_2 \in I$, where $x_1 < x_2$. since $[x_1, x_2] \subseteq I$ and $(x_1, x_2) \subseteq I$, so
$g(x)$ is continuous in $[x_1, x_2]$ and differentiable in (x_1, x_2).

Therefore, by Lagrange's Mean Value Theorem (4.11), there exists a point c, $x_1 < c < x_2$ such that

$$\frac{g(x_1) - g(x_2)}{x_1 - x_2} = g'(c)$$

or,

$$g(x_1) - g(x_2) = (x_1 - x_2)g'(c) \tag{126}$$

Also it is given that $f(x) = x + \epsilon g(x)$, so

$$f(x_1) = x_1 + \epsilon g(x_1) \text{ and } f(x_2) = x_2 + \epsilon g(x_2)$$

Subtracting, we get

$$\begin{aligned}
f(x_1) &- f(x_2) \\
&= (x_1 - x_2) + \epsilon(g(x_1) - g(x_2)) \\
&= (x_1 - x_2) + \epsilon(x_1 - x_2)g'(c) \text{ (by 126)} \\
&= (x_1 - x_2)(1 + \epsilon g'(c)). \text{ Thus}
\end{aligned}$$

$f(x_1) - f(x_2) = (x_1 - x_2)(1 + \epsilon g'(c))$, $x_1 < c < x_2$, which implies that

$$|f(x_1) - f(x_2)| = |(x_1 - x_2)||1 + \epsilon g'(c)| \tag{127}$$

Now

$$|1 + \epsilon g'(c)| \geq 1 - |\epsilon g'(c)|$$

$$= 1 - \epsilon|g'(c)|$$

$\geq 1 - \epsilon m$, (because by hypothesis $|g'(x)| \leq m$ for all $x' \in I$ and so $|g'(c)| \leq m$ or $-|g'(c)| \geq -m$).

Thus, $|1 + \epsilon g'(c)| \geq 1 - \epsilon m$.

Hence $|1 + \epsilon g'(c)| > 0$ if $1 - \epsilon m > 0$

i.e. if $1 > \epsilon m$

i.e. if $\epsilon m < 1$

i.e. if $\epsilon < \frac{1}{m}$

So if $0 < \epsilon < \frac{1}{m}$ then $|1 + \epsilon g'(c)| > 0$.

Therefore, with this choice of ϵ, it follows from (127) that

$$|f(x_1) - f(x_2)| = 0 \text{ if and only if } |x_1 - x_2| = 0$$
$$\text{i.e. } f(x_1) - f(x_2) = 0 \text{ if and only if } x_1 - x_2 = 0$$
$$\text{i.e. } f(x_1) = f(x_2) \text{ if and only if } x_1 = x_2$$

This shows that if $0 < \epsilon < \frac{1}{m}$, then f is one - one.

Exercise 4.32. Find θ of Lagrange's Mean Value Theorem (4.11) for $f(x) = \alpha x^2 + \beta x + \gamma$ in $[a, b]$.

Solution: Applying Lagrange's Mean Value Theorem (4.11) to $f(x)$ in $[a, b]$ where

$f(x) = \alpha x^2 + \beta x + \gamma$, we get

$$f(b) - f(a) = (b - a)f'(a + \theta(b - a)) \tag{128}$$

But
$$f(b) = \alpha b^2 + \beta b + \gamma \tag{129}$$

and
$$f(a) = \alpha a^2 + \beta b + \gamma \tag{130}$$

Also, $f'(x) = 2\alpha x + \beta$

so

$$f'(a + \theta(b - a)) = 2\alpha(a + \theta(b - a)) + \beta \qquad (131)$$

Using (129), (130) and (131) in (128), we get

$$\alpha b^2 + \beta b + \gamma - (\alpha a^2 + \beta b + \gamma) = (b - a).(2\alpha\ (a + \theta(b - a)) + \beta)$$

$$\Rightarrow \alpha(b^2 - a^2) + \beta(b - a) = (b - a).(2\alpha a + 2\alpha\theta(b - a) + \beta)$$

$$\Rightarrow (b - a)[\alpha(b + a) + \beta] = (b - a).(2\alpha a + 2\alpha(b - a)\theta + \beta)$$

$$\Rightarrow \alpha(b + a) = 2\alpha a + 2\alpha(b - a))\theta$$

$$\Rightarrow \alpha(b - a) = 2\alpha(b - a)\theta$$

$$\Rightarrow 1 = 2\theta$$

$$\Rightarrow \theta = \tfrac{1}{2}$$

i.e. $\theta = \tfrac{1}{2}$.

Exercise 4.33. Suppose $f'(x)$ exists finitely for $-\infty < x < \infty$ and $f'(x) \neq 1$ for $-\infty < x < \infty$. Prove that $f(x)$ has at most one fixed point.

Solution: It is given that $f'(x)$ exists finitely for $-\infty < x < \infty$ and $f'(x) \neq 1$ for $-\infty < x < \infty$. If possible suppose $f(x)$ has two fixed points x_1 and x_2 say $x_1 < x_2$, then $f(x_1) = x_1$ and $f(x_2) = x_2$.

Since by hypothesis $f'(x)$ exists finitely for $-\infty < x < \infty$, therefore, $f(x)$ exists finitely for $x \in [x_1, x_2]$ and so Lagrange's Mean Value Theorem (4.11), implies that $\frac{f(x_1) - f(x_2)}{x_1 - x_2} = f'(c)$, for some c, $x_1 < c < x_2$

or $f(x_1) - f(x_2) = (x_1 - x_2)f'(c)$, $x_1 < c < x_2$

or $(x_1 - x_2) = (x_1 - x_2)f'(c)$, $x_1 < c < x_2$

i.e. $f'(c) = 1$ for some c, $x_1 < c < x_2$ (because $x_1 - x_2 \neq 0$)

or $f'(c) = 1$, for some c, $\infty < c < \infty$,

which contradicts the hypothesis because it is given that $f'(x) \neq 1$, for $\infty < x < \infty$. Hence our supposition that $f(x)$ has two fixed points must be wrong.

It follows that $f(x)$ has at most one fixed point.

Exercise 4.34. Deduce from Lagrange's Mean Value Theorem (4.11), that if $f'(x) > 0$ for $a < x < b$, then $f(x)$ is increasing and if $f'(x) < 0$ for $a < x < b$ then $f(x)$ is decreasing and if $f'(x) = 0$ for $a < x < b$, then $f(x)$ is constant.

Solution: Let $a < x_1 < x_2 < b$.

Applying Lagrange's Mean Value Theorem (4.11), to $f(x)$ in $[x_1, x_2]$, we get

$$f(x_2) - f(x_1) = (x_2 - x_1)f'(c) \tag{132}$$

for some c, $x_1 < c < x_2$

If $f'(x) > 0$ for $a < x < b$.
Then, as $a < c < b$, and so $f'(c) > 0$.

Therefore, from (132), it follows that

$$f(x_2) - f(x_1) > 0 \text{ or } f(x_2) > f(x_1).$$

Thus for $x_2 > x_1$, we have $f(x_2) > f(x_1)$, which shows that $f(x)$ is an increasing function.

Now if $f'(x) < 0$ for $a < x < b$.
Then, as $a < c < b$ and so $f'(c) < 0$.

Therefore, from (132), it follows that

$$f(x_2) - f(x_1) < 0 \text{ (because } x_2 - x_1 > 0 \text{ as } x_2 > x_1)$$

$$\text{or } f(x_2) < f(x_1).$$

Thus for $x_2 > x_1$, we have $f(x_2) < f(x_1)$, which shows that $f(x)$ is an decreasing function.

Finally If $f'(x) = 0$ for $a < x < b$.
Then, as $a < c < b$, so $f'(c) = 0$.

Therefore, from 132, it follows that

$$f(x_2) - f(x_1) = 0 \text{ or } f(x_2) = f(x_1) \ .$$

Thus for $x_2 > x_1$, we have $f(x_2) = f(x_1)$, which shows that $f(x)$ is an constant function.
Hence the result

Note:- If $f(x) = -x$, then for $x = 1$, $f(x) = -1$; $x = 2$, $f(x) = -2$ and when $f(x) = x$, then for $x = 1$, $f(x) = 1$; $x = 2$, $f(x) = 2$ i.e. if $x_1 > x_2$, then $f(x_1) < f(x_2)$.

Exercise 4.35. Suppose $-1 < x < 0$, prove $1 + \frac{x}{2\sqrt{1+x}} < \sqrt{1+x} < 1 + \frac{x}{2}$.

Solution: Suppose $-1 < x < 0$.
Consider $f(x) = \sqrt{1+x}$, then $f'(x) = \frac{1}{2\sqrt{1+x}}$, which exists at $x \neq -1$.

Applying Lagrange's Mean Value Theorem to (4.11) $f(x)$, we get

$$f(x) = f(0) + (x - 0)f'(0 + \theta x), \ 0 < \theta < 1$$

or, $f(x) = 1 + xf'(\theta x), \ 0 < \theta < 1$

so for $0 < \theta < 1$;

$$f(x) = 1 + \frac{x}{2\sqrt{1 + \theta x}} \tag{133}$$

Now $0 < \theta < 1, -1 < x < 0$ implies that $\theta x < 0$. So $1 + \theta x < 1$ i.e $\sqrt{1 + \theta x} < 1$ or $\frac{1}{\sqrt{1+\theta x}} > 1$; multiplying both sides by x we get,

$\frac{x}{\sqrt{1+\theta x}} < x$ (because $-1 < x < 0$).

Therefore, (133) implies that

$$f(x) < 1 + \frac{x}{2} \qquad (134)$$

Also $\theta x > x$ implies that

$1 + \theta x > 1 + x$

or $\sqrt{1 + \theta x} > \sqrt{1 + x}$

or $\frac{1}{\sqrt{1+\theta x}} < \frac{1}{\sqrt{1+x}}$

or $\frac{x}{\sqrt{1+\theta x}} < \frac{x}{\sqrt{1+x}}$.

Therefore (133) implies that

$$f(x) > 1 + \frac{x}{2\sqrt{1+x}}. \qquad (135)$$

Now combining (134) and (135), we get

$1 + \frac{x}{2\sqrt{1+x}} < f(x) < 1 + \frac{x}{2}$

or $1 + \frac{x}{2\sqrt{1+x}} < \sqrt{1+x} < 1 + \frac{x}{2}$

Exercise 4.36. Suppose $f(x)$ is continuous and finite for finite x. Further $f(x + y) = f(x) + f(y)$. Then prove that $f(x) = cx$ where c is a constant.

Solution: It is given that $f(x)$ is continuous and finite for finite x and $f(x+y) = f(x)+f(y)$. Taking $x = y = 0$, it follows from $f(x+y) = f(x)+f(y)$ that $f(0+0) = f(0)+f(0)$ i.e. $f(0)+f(0)-f(0) = 0$ implies that $f(0) = 0$.

Now let $f(1) = c$, then $f(2) = f(1+1) = f(1) + f(1) = c + c = 2c$.

Now $f(2) = 2c$ and $f(3) = f(2+1) = f(2) + f(1) = 2c + c = 3c$.
Similarly if n is any positive integer then $f(n) = nc$. So the result is proved in this case.
Let p and q be any positive integers. Then

$$cp = f(p)$$

$$= f(\tfrac{p}{q} + \tfrac{p}{q} + \dots + \tfrac{p}{q}) \ (\text{------}q \text{ times------})$$

$$= f(\tfrac{p}{q}) + f(\tfrac{p}{q}) + \dots + f(\tfrac{p}{q})$$

$$= qf(\tfrac{p}{q}).$$

Thus $cp = qf(\tfrac{p}{q})$, or $f(\tfrac{p}{q}) = c\tfrac{p}{q}$

Thus if x is a positive rational number then, $f(x) = cx$ and so the result is proved in this case.

Let α be any positive irrational number. Then there exists a sequence $\{\tfrac{p_n}{q_n}\}_1^\infty$ of irrational number (by Rational Density Theorem (8.4)) such that

$$lim_{n\to\infty} \tfrac{p_n}{q_n} = \alpha$$

Since by hypothesis $f(x)$ is continuous it follows from above that

$$lim_{n\to\infty} f(\tfrac{p_n}{q_n}) = f(\alpha)$$

which implies that

$$lim_{n\to\infty} c(\tfrac{p_n}{q_n}) = f(lim_{n\to\infty} \tfrac{p_n}{q_n})$$

or $clim_{n\to\infty} \tfrac{p_n}{q_n} = f(\alpha)$

or $c\alpha = f(\alpha)$ [because $lim_{n\to\infty} \tfrac{p_n}{q_n} = \alpha$]

This shows that result is true in case x is a positive irrational number. Thus we have proved that $f(x) = cx$, $x \geq 0$
because $f(x + y) = f(x) + f(y)$; taking $x = -y$, we get $f(0) = f(x) + f(-x)$

or $0 = f(x) + f(-x)$

or $f(x) = f(-x)$.

Therefore from this, it follows that $f(x) = cx$, for all x.

Exercise 4.37. Let $f''(x)$ exists in $[0, l]$ and

$$M_0 = Max|f(x)|, M_1 = Max|f'(x)|, M_2 = Max|f''(x)| \text{ in } [0, l] \text{ then}$$

$$M_1 \leq 2\sqrt{M_0 M_2} \text{ if } l \geq 2\sqrt{\frac{M_0}{M_2}}.$$

Give the example to show that constant 2 is the best possible.

Solution: It is given that $f'(x)$ exists in $[0, l]$ and

$$M_0 = Max|f(x)|, M_1 = Max|f'(x)|, M_2 = Max|f''(x)| \text{ in } [0, l].$$
Therefore, applying Taylor's Theorem (4.14) to $f(x)$ we get

$$f(0) = f(x) + (0 - x)f'(x) + \frac{(0 - x)^2}{2!}f''(c_1) \tag{136}$$

for some c_1; $0 < c_1 < x$

and
$$f(l) = f(x) + (l - x)f'(x) + \frac{(l - x)^2}{2!}f''(c_2) \tag{137}$$

for some c_2; $l < c_2 < 0$

Subtracting (136) from (137), we get

$$f(l) - f(0) = lf'(x) + \frac{(l-x)^2}{2!}f''(c_2) - \frac{x^2}{2!}f''(c_1)$$

or $lf'(x) = f(l) - f(0) - \frac{(l-x)^2}{2!}f''(c_2) + \frac{x^2}{2!}f''(c_1)$

or $f'(x) = \frac{f(l)}{l} - \frac{f(0)}{l} - \frac{(l-x)^2}{2l}f''(c_2) + \frac{x^2}{2l}f''(c_1)$

Taking modulus of both sides, we get

$$|f'(x)| = \left|\frac{f(l)}{l} - \frac{f(0)}{l} - \frac{(l-x)^2}{2l}f''(c_2) + \frac{x^2}{2l}f''(c_1)\right|$$

$$\text{or, } |f'(x)| \le \frac{|f(l)|}{l} + \frac{|f(0)|}{l} + \frac{(l-x)^2}{2l}|f''(c_2)| + \frac{x^2}{2l}|f''(c_1)|$$

$$\text{or, } |f'(x)| \le \frac{M_0}{l} + \frac{M_0}{l} + \frac{(l-x)^2}{2l}M_2 + \frac{x^2}{2l}M_2$$

$$\text{or, } |f'(x)| = \frac{2M_0}{l} + \frac{M_2}{2l}(x^2 + (l-x)^2)$$

$$\text{or, } |f'(x)| \le \frac{2M_0}{l} + \frac{M_2}{2l}l^2 \text{ (because } (x^2 + (l-x)^2) \le l^2)$$

Therefore, $|f'(x)| \le \frac{2M_0}{l} + \frac{M_2 l}{2}$.

Since L.H.S is independent of l, therefore,

$$|f'(x)| \le Minimum\left(\frac{2M_0}{l} + \frac{M_2 l}{2}\right) \text{ with respect to } l.$$

But minimum of $\left(\frac{2M_0}{l} + \frac{M_2 l}{2}\right)$ is attained for $l^2 = \frac{4M_0}{M^2}$ or $l = 2\sqrt{\frac{M_0}{M_2}}$ and the required minimum is

$$\frac{2M_0}{2\sqrt{\frac{M_0}{M_2}}} + \frac{M_2}{2}.2\sqrt{\frac{M_0}{M_2}} = 2\sqrt{M_0 M_2}.$$

So if $l \ge 2\sqrt{\frac{M_0}{M_2}}$ then $|f'(x)| \le 2\sqrt{M_0 M_2}$

Therefore, $Max|f'(x)| \le 2\sqrt{\frac{M_0}{M_2}}, x \in [0, l]$ or $M_1 \le 2\sqrt{M_0 M_2}$.

Thus $M_1 \le 2\sqrt{M_0 M_2}$ if $l \ge \sqrt{\frac{M_0}{M_2}}$.

Now to see that constant 2 is the best possible, consider $f(x) = x^2 - 2, x \in [0, 2]$. Here $l = 2$ and $Max|f(x)| = M_0 = 2, x \in [0, 2]$.

Also $|f'(x)| = 2x$ and so $Max|f'(x)| = M_1 = 4, x \in [0, 2]$ and $f''(x) = 2$ and $Max|f''(x)| = M_2 = 2, x \in [0, 2]$.

Therefore, $2\sqrt{M_0 M_2} = 2\sqrt{\frac{2}{2}} = 2$ and $M_1 = 2\sqrt{M_0 M_2}$ (because $M_1 = 4, M_0 = M_2 = 2$).

Exercise 4.38. Suppose $f'(x)$ exists finitely. Prove that

$$\phi(h, k) = \frac{f(a+h) - f(a-k)}{h+k} - f'(a) \to 0$$

as h and k tends to zero simultaneously through positive values. Also prove that h and k tend to zero through positive values can be relaxed if $f(x)$ possesses a continuous derivative in the neighborhood of a.

Solution: It is given that

$$\phi(h, k) = \frac{f(a+h) - f(a-k)}{h+k} - f'(a)$$

or $\phi(h, k) = \frac{f(a+h)}{h+k} - \frac{f(a-k)}{h+k} - \frac{h}{h+k} f'(a) - \frac{k}{h+k} f'(a)$

$$= \left[\frac{f(a+h)}{h+k} - \frac{f(a-k)}{h+k}\right] - \left[\frac{h}{h+k} f'(a) + \frac{k}{h+k} f'(a)\right]$$

$$= \frac{h}{h+k}\left[\frac{f(a+h)}{h} - f'(a)\right] - \frac{k}{h+k}\left[\frac{f(a-k)}{k} + f'(a)\right]$$

$$= \frac{h}{h+k}\left[\frac{f(a+h) - f(a)}{h} - f'(a)\right] - \frac{k}{h+k}\left[\frac{f(a-k) - f(a)}{k} + f'(a)\right]$$

$$\phi(h, k) = \frac{h}{h+k}\left[\frac{f(a+h) - f(a)}{h} - f'(a)\right] + \frac{k}{h+k}\left[\frac{f(a-k) - f(a)}{-k} - f'(a)\right]$$

$$\tag{138}$$

Now if $h > 0$ and $k > 0$ then $0 < \frac{k}{h+k} < 1$ and $0 < \frac{h}{h+k} < 1$. Taking modulus of both sides in (138), we get

$$|\phi(h, k)| = \left|\frac{h}{h+k}\left[\frac{f(a+h) - f(a)}{h} - f'(a)\right] + \frac{k}{h+k}\left[\frac{f(a-k) - f(a)}{-k} - f'(a)\right]\right|$$

or, $|\phi(h, k)| \leq \frac{h}{h+k}\left|\frac{f(a+h) - f(a)}{h} - f'(a)\right| + \frac{k}{h+k}\left|\frac{f(a-k) - f(a)}{-k} - f'(a)\right|$

Now for $h > 0$ and $k > 0$, $0 < \frac{k}{h+k} < 1$ and $0 < \frac{h}{h+k} < 1$. Therefore,

$$|\phi(h,k)| \le |\frac{f(a+h) - f(a)}{h} - f'(a)| + \frac{f(a-k) - f(a)}{-k} - f'(a)| \quad (139)$$

Now $f'(a)$ exists finitely, therefore,

$$lim_{h\to 0+}\frac{f(a+h)-f(a)}{h} = f'(a) = lim_{k\to 0+}\frac{f(a-k)-f(a)}{-k}.$$

Letting h and k tend to zero through positive values, it follows from (139), that $\phi(h,k)| \le 0$ which implies that

$$lim_{h\to 0+,k\to 0+}\phi(h,k) = 0.$$

Hence $\phi(h,k)$ approach zero as h and k approach zero simultaneously through positive values.

In the second case applying Lagrange's Mean Value Theorem (4.11) to $f(x)$, we get

$$\frac{f(a+h)-f(a-k)}{(a+h)-(a-k)} = f'[(a-k) + (h+k)\theta], \ 0 < \theta < 1$$

or,

$$\frac{f(a+h) - f(a-k)}{(h-k)} = f'[(a-k) + (h+k)\theta] \quad (140)$$

Now if the derivative of $f(x)$ is continuous in the neighborhood of $'a'$, then $(a-k) + (h+k)\theta \to 0$ as $h, k \to 0$ in such a way that $h + k \ne 0$. Now

$$f'[(a-k) + (h+k)\theta] \to f'(a), \text{ and using this in (141), we get}$$

$$\frac{f(a+h)-f(a-k)}{h+k} \to f'(a) \text{ as } h, k \to 0, \text{ therefore}$$

$$\phi(h,k) = \frac{f(a+h)-f(a-k)}{h+k} - f'(a) \to 0 \text{ as } h, k > 0.$$

From this it follows that the condition that h and k should tend to zero simultaneously through positive values can be relaxed if $f(x)$ has continuous derivative in the neighborhood of a'. This proves the result completely.

Theorem 4.39. Existence of Logarithm:

Suppose $x > 0$ and $b > 1$, then there is one and only one real number y such that $b^y = x$ or $logb^x = y$.

Proof. Let there be one and only one real number y such that $b^y = x$. It follows from the fact that if there are two real numbers $y_1, y_2; y_1 \neq y_2$, then $b^{y_1} \neq b^{y_2}$ and their common value is not equal to x. Now let $x > 0$ and $b > 1$. We now prove the existence of y such that $b^y = x$.

Let $1 < x < b$. Consider $f(\lambda) = b^\lambda - x$, in $[0, 1]$. Then $f(x)$ is continuous in $[0, 1]$. Further $f(0) = 1 - x < 0$, $f(1) = b - x > 0$.

Thus $f(\lambda)$ is continuous in $[0, 1]$, $f(0)$ and $f(1)$ differ in sign, therefore, there exists a real number y, $0 < y < 1$ such that $f(y) = 0$ i.e. $b^y - x = 0$ or $b^y = x$.

Now suppose $x = 1$, then $y = 0$ and if $x = b$, then $y = 1$, so that in these two cases there is nothing to prove.

Suppose $x > b$ $(b > 1)$, then $b^\lambda \to \infty$ as $\lambda \to \infty$.

Therefore, we can choose a sufficiently large $\lambda = k$(say) such that

$$b^k > x. \tag{141}$$

Now consider $f(\lambda) = b^\lambda - x$, in $[1, k]$. Then $f(\lambda)$ is continuous in $[1, k]$.

Also $f(1) = b - x < 0$ (because $x > b$) and $f(k) = b^k - x > 0$ (using (141)). Thus $f(\lambda)$ is continuous in $[1, k]$, now $f(1)$ and $f(k)$ differ in sign, therefore, there exists a y, $1 < y < k$ such that $f(y) = 0$ or $b^y - x = 0$ or $b^y = x$.

Now let $0 < x < 1$. Then $b^\lambda \to 0$ as $\lambda \to -\infty$, therefore, we can choose $\lambda = k'$; $k' < 0$ such that

$$b^{k'} < x \tag{142}$$

Consider $f(\lambda) = b^\lambda - x$ in $[k', 0]$, then $f(k') = b^{k'} - x < 0$ (by using (142) and $f(0) = 1 - x > 0$.

Thus $f(\lambda)$ is continuous in $[k', 0]$. Now $f(k')$ and $f(0)$ differ in sign, therefore, there exist a point y, $k' < y < 0$ such that $f(y) = 0$ or $b^y - x = 0$, i.e. $b^y = x$.

This proves the result completely. $\square$

Exercise 4.40. Prove that $sinx < x$ for $0 < x < \frac{\pi}{2}$.

Solution: Consider $f(x) = x - sinx$, then $f(x)$ is continuous in $[0, \frac{\pi}{2}]$.

Let $0 < x < \frac{\pi}{2}$. Then $f'(x) = 1 - cosx$.

But for $0 < x < \frac{\pi}{2}$, $1 - cosx > 0$, and therefore, for $0 < x < \frac{\pi}{2}$, $f'(x) > 0$.

This shows that $f(x)$ is an increasing function. Therefore, $x > 0$ implies that $f(x) > f(0)$, i.e. $x - sinx > 0$, i.e. $x > sinx$, i.e. $sinx < x$ for $0 < x < \frac{\pi}{2}$, which proves the result.

Definition 4.41. Suppose $f(x)$ is defined at $x = x_0$. If $lim_{h \to 0} f(x_0 \pm h)$ exists, but their common value is not equal to $f(x_0)$, then $f(x)$ is said to have a **simple discontinuity or discontinuity of first kind** at the point x_0.
If one or both of these limits fail to exist or are infinite then $f(x)$ is said to have a **discontinuity of second kind at x_0** .

Exercise 4.42. Prove that if $f'(x)$ exists finitely in $[a, b] = I$, then it cannot have a discontinuity of first kind.

Solution: It is given that $f'(x)$ exists finitely in $[a, b] = I$.

Let $x_0 \in [a, b]$. Then $f'(x_0)$ exists. If h is a small positive number, then $f'(x_0) = lim_{h \to 0} \frac{f(x_0 + h) - f(x_0)}{h}$

or $f'(x_0) = lim_{h \to 0} f'(x_0 + \theta h), \ 0 < \theta < 1 = f'(x_0^+)$.

If right hand limit exists finitely, then it is necessarily equal to $f'(x_0)$.

Also, $f'(x_0) = lim_{h \to 0} \frac{f(x_0 - h) - f(x_0)}{-h}$

or $f'(x_0) = lim_{h \to 0} f'(x_0 - \theta h), \ 0 < \theta < 1 = f'(x_0^-)$.

Thus if right hand limit and left hand limit exist finitely, then both are equal to $f'(x_0)$.

Hence $f'(x)$ can not have a discontinuity of first kind.

Exercise 4.43. Let

$$f(x) = -1, -1 \le x \le 0; \ f(x) = 1, 0 < x \le 1.$$

Is there a function $g(x)$ such that $g'(x) = f(x)$ in $[-1, 1]$.

Solution: It is given that

$$f(x) = -1, -1 \le x \le 0; \ f(x) = 1, 0 < x \le 1.$$

We note that $f(x)$ has a simple discontinuity at $x = 0$. In fact $f(0) = -1$. Also for $h > 0$, $f(0 + h) = 1$, therefore, $lim_{h \to 0} f(0 + h) = 1$ and $f(0 - h) = -1$, therefore, $lim_{h \to 0} f(0 - h) = -1$.

Thus $lim_{h \to 0} f(0 + h)$ and $lim_{h \to 0} f(0 - h)$ exist but their common value is not equal to $f(0)$.

This shows that $f(x)$ has a simple discontinuity or discontinuity of first kind at $x = 0$.

Now if there exist a function $g(x)$ such that $g'(x) = f(x)$ in $[-1, 1]$, then it follows that $g'(x)$ exists finitely in $[-1, 1]$ and has a simple discontinuity at $x = 0$ (because $g'(x) = f(x)$ in $[-1, 1]$ and we have proved that $f(x)$ has a simple discontinuity at $x = 0$ which cannot happen by the above result (4.42)).

Hence no such function $g(x)$, such that $g'(x) = f(x)$ in $[-1, 1]$ can exist.

EXERCISES

(1) If $f(x) = sinx\,sin(\frac{1}{sinx})$, when $0 < x < \pi$ and $\pi < x < 2\pi$;
$f(x) = 0$, when $x = 0, \pi, 2\pi$;

then show that $f(x)$ is continuous in $[0, 2\pi]$ but derivable everywhere except at $x = 0, \pi$ and 2π.

(2) Let $f(x) = x\,tan^{-1}\frac{1}{x}$ for $x \neq 0$ and $f(0) = 0$, then find the left and right derivatives of $f(x)$ at $x = 0$. Does $f'(0)$ exist.

(3) Let $f(x) = 1 - x^2$, when $0 < x \leq 1$
$f(x) = logx$ when $1 < x \leq 2$
$f(x) = log2 - 1 + \frac{x}{2}$ when $2 < x < \infty$.

Obtain the derivative of the function.

(4) Let $f(x) = e^{\frac{-1}{x^2}} sin\frac{1}{x}$ for $x \neq 0$ and $f(0) = 0$. Is $f(x)$ derivable?

(5) Verify Rolle's Theorem for the following functions:

    (a) $f(x) = sinx\,cosx$ on $[0, \frac{\pi}{2}]$

    (b) $f(x) = 4sinx$ on $[0, \pi]$

       iii) $f(x) = log(sinx)$ on $[\frac{\pi}{6}, \frac{5\pi}{6}]$

(6) Verify Lagrange's Mean Value Theorem for the following functions:

    (a) $f(x) = \sqrt{x}$ on $[4, 9]$

    (b) $f(x) = ax^2 + bx + c$ on $[\alpha, \beta]$

    (c) $f(x) = (x - 1)(x - 2)(x - 3)$ on $[5, 7]$

(7) Verify Cauchy Mean Value Theorem for the following functions:

    (a) $f(x) = sinx$, $g(x) = cosx$ on $[0, \frac{\pi}{2}]$

    (b) $f(x) = x^2$, $g(x) = x$ on $[a, b]$

    (c) $f(x) = \frac{1}{x^2}$, $g(x) = \frac{1}{x}$ on $[1, 2]$

(8) If $f(x) = xcos\frac{1}{x}$, $x \neq 0$ and $f(0) = 0$ defined on $[-1, 1]$, then test whether Lagrange's Mean Value Theorem holds for $f(x)$.

(9) Use Mean Value Theorem for

    $f(h) = f(0) + hf'(\theta h), 0 < \theta < 1$ to prove

(a) $lim_{h\to 0+0}\theta = \frac{1}{x}$ if $f(x) = cosx$

(b) $lim_{h\to 0+0}\theta = \frac{1}{\sqrt{3}}$ if $f(x) = sinx$

(10) Show that

(a) $logx = (x-1) - \frac{1}{2}(x-1)^2 + \frac{1}{3}(x-1)^3 - \ldots$ for $0 < x \le 2$

(b) $\frac{1}{x} = \frac{1}{2} - \frac{1}{2^2}(x-2) + \frac{1}{2^3}(x-2)^2 - \ldots$ for $0 < x < 4$

Chapter5

INEQUALITIES

The Mean Value theorem is helpful in obtaining functional inequalities. Sometimes from one established functional inequality we are able to deduce different form of functional inequality and in that case two such inequalities are said to be equivalent. The functional inequalities play important role in quantitative analysis.

Definition 5.1. If $a_1, a_2, a_3, ..., a_n$ are integers, then their Arithmetic mean is given by

$$A = \frac{a_1 + a_2 + a_3 + ... + a_n}{n}.$$

Definition 5.2. If $a_1, a_2, a_3, ..., a_n$ are integers, then their Geometric mean is given by

$$G = (a_1.a_2.a_3...a_n)^{\frac{1}{n}}.$$

Definition 5.3. If $a_1, a_2, a_3, ..., a_n$ are positive integers, then their Harmonic mean is given by

$$H = \frac{n}{\frac{1}{a_1} + \frac{1}{a_2} + \frac{1}{a_3} + ... + \frac{1}{a_n}}$$

Theorem 5.4. *If $a_1, a_2, a_3, ..., a_n$ are positive real numbers, then $A \geq G \geq H$ and the sign of equality holds when $a_1 = a_2 = a_3 = ... = a_n$.*

Proof. Let $a_1, a_2, a_3, ..., a_n$ be positive real numbers. Now

$$0 \leq (\sqrt{a_1} - \sqrt{a_2})^2 = a_1 + a_2 - 2\sqrt{a_1 a_2}. \text{ Thus } a_1 + a_2 - 2\sqrt{a_1 a_2} \geq 0$$

or $a_1 + a_2 \geq 2\sqrt{a_1 a_2}$

or

$$\frac{a_1 + a_2}{2} \geq \sqrt{a_1 a_2} = (a_1 a_2)^{\frac{1}{2}} \tag{143}$$

and equality holds when $a_1 = a_2$.

This shows that result is true for $n = 2$. Now Consider a_1, a_2, a_3, a_4 as positive reals. From (143), it follows that

$$\frac{a_3 + a_4}{2} \geq \sqrt{a_3 a_4} \tag{144}$$

Therefore,

$$\frac{a_1+a_2+a_3+a_4}{4} = \frac{\frac{a_1+a_2}{2} + \frac{a_3+a_4}{2}}{2} \geq \frac{\sqrt{a_1+a_2}+\sqrt{a_3+a_4}}{2} \text{ (by using (143) and}$$
(144)).

Thus $\frac{a_1+a_2+a_3+a_4}{4} \geq \frac{a_1+a_2}{2} + \frac{a_3+a_4}{2} \geq (\sqrt{a_1 a_2}\sqrt{a_3 a_4})^{\frac{1}{2}}$,

Therefore, $\frac{a_1+a_2+a_3+a_4}{4} \geq (a_1 a_2 a_3 a_4)^{\frac{1}{4}}$ and equality holds when $a_1 = a_2 = a_3 = a_4$.

This shows that result is true for $n = 4 = 2^2$.

Similarly it can be proved that the result is true for $n = 2^3$ and for any positive integral power of 2, say 2^m.

We choose the positive integer n in such a way that $2^m > n$,i.e. $2^m - n$ is a positive integer. Let $k = \frac{a_1+a_2+a_3+...+a_n}{n}$. Consider 2^m numbers $a_1, a_2, a_3, ..., a_n, k, k, ..., k$.

Hence the number of $k's$ is $2^m - n$.

Now consider the product $a_1.a_2.a_3...a_n.k.k....k$. Then there are 2^m numbers in this product where n is a positive integer not of the form 2^m. Since we have already proved the result for integral powers of 2. Therefore, the result is true for 2^m also.So,

$$(a_1.a_2.a_3...a_n.k.k....k)^{\frac{1}{2m}} \leq \frac{a_1+a_2+a_3...+a_n+k+k+...+k}{2^m}.$$

or,

$$(a_1.a_2.a_3...a_n.k^{2^m-n})^{\frac{1}{2m}} \leq \frac{nk+(2^m-n)k}{2m}.$$

So,

$$(a_1.a_2.a_3...a_n.k^{2^m-n})^{\frac{1}{2m}} \leq k.$$

This gives $a_1.a_2.a_3...a_n.k^{2^m-n} \leq k^{2^m}$

i.e, $a_1.a_2.a_3...a_n.k^{2^m}k^{-n} \leq k^{2m}$

i.e, $a_1 a_2...a_n \frac{k^{2m}}{k^{2m}}\frac{1}{k^n} \leq \frac{k^{2m}}{k^{2m}}$

i.e, $a_1.a_2.a_3...a_n \frac{1}{k^n} \leq 1$

i.e, $a_1.a_2.a_3...a_n \leq k^n$

i.e, $(a_1.a_2.a_3...a_n)^{\frac{1}{n}} \leq k$

or,

$$(a_1.a_2.a_3...a_n)^{\frac{1}{n}} \leq \frac{a_1 + a_2 + a_3 + ... + a_n}{n} \tag{145}$$

(because $k = \frac{a_1+a_2+a_3+...+a_n}{n}$) and equality hold if $a_1 = a_2 = ... = a_n$. Therefore, $G \leq A$.

Now replace $a_i, i = 1, 2, 3, ..., a_n$ by $\frac{1}{a_i}$ in (145), we get

$$(\frac{1}{a_1} + \frac{1}{a_2} + \frac{1}{a_3} + ... + \frac{1}{a_n})^{\frac{1}{n}} \leq \frac{\frac{1}{a_1}+\frac{1}{a_2}+\frac{1}{a_3}+...+\frac{1}{a_n}}{n}$$

or, $\dfrac{1}{(a_1 a_2 ... a_n)^{\frac{1}{n}}} \leq \dfrac{\frac{1}{a_1} + \frac{1}{a_2} + \frac{1}{a_3} + ... + \frac{1}{a_n}}{n}$

or, $(a_1 a_2 ... a_n)^{\frac{1}{n}} \geq \dfrac{n}{\frac{1}{a_2} + \frac{1}{a_3} + ... + \frac{1}{a_n}}$

and equality holds if $a_1 = a_2 = ... = a_n$, therefore, $G \geq H$.

Hence it follows that $A \geq G \geq H$ and equality holds if all the $a_i's$ are equal. $\square$

Corollary 5.5. *If all the a's are positive, prove that* $\sum_{r=1}^{n} a_r \sum_{r=1}^{n} \frac{1}{a^r} \geq n^2$.

Proof. From above result $A \geq G \geq H$, therefore,

$\dfrac{a_1 + a_2 + a_3 ... + a_n}{n} \geq \dfrac{n}{\frac{1}{a_1} + \frac{1}{a_2} + \frac{1}{a_3} + ... + \frac{1}{a_n}}$

or, $\dfrac{\sum_{r=1}^{n} a_r}{n} = \dfrac{n}{\sum_{r=1}^{n} \frac{1}{a^r}}$.

This gives $\sum_{r=1}^{n} a_r \sum_{r=1}^{n} \frac{1}{a^r} \geq n^2$. $\square$

Theorem 5.6. **(Cauchy Schwartz Inequality)**: *If* a_i, b_i $(i = 1, 2, 3, ..., n)$ *are real numbers, then*

$$\sum_{i=1}^{n} a_i b_i \leq \left(\sum_{i=1}^{n} a_i^2\right)^{\frac{1}{2}} \left(\sum_{i=1}^{n} b_i^2\right)^{\frac{1}{2}}$$

and sign of equality hold if

$a_i \lambda + b_i = 0$, $i = 1, 2, 3, ..., n$; *where* λ *is a real number.*

Proof. It is given that a_i, b_i $(i = 1, 2, 3, ..., n)$ are real numbers. Let λ be any real number. Then for $i = 1, 2, ..., n$, $(a_i \lambda + b_i)^2 \geq 0$,

i.e, $a_i^2 \lambda^2 + 2 a_i b_i \lambda + b_i^2 \geq 0$, where equality holds if $a_i \lambda + b_i = 0$.

Adding these n inequalities we get

$$(\textstyle\sum_{i=1}^{n} a_i^2)\lambda^2 + 2(\sum_{i=1}^{n} a_i b_i) + (\sum_{i=1}^{n} b_i^2)^{\frac{1}{2}} \geq 0$$

or,

$$A\lambda^2 + 2B\lambda + C \geq 0 \tag{146}$$

where $A = \sum_{i=1}^{n} a_i^2$, $B = \sum_{i=1}^{n} a_i b_i$ and $c = \sum_{i=1}^{n} b_i^2$.

If $A = 0$, then all the $a's$ are equal to zero and L.H.S and R.H.S of the given inequality are both equal to zero. So that there is nothing to prove in this case.

So, let $A > 0$. Consider $f(\lambda) = A\lambda^2 + 2B\lambda + C$. Then $f(\lambda) \geq 0$ by (146), for all real λ.

We claim that $B^2 \leq AC$.

If possible suppose $B^2 > AC$. Now
$f(\frac{-B}{A})$
$= A(\frac{-B}{A})^2 + 2B(\frac{-B}{A}) + C$
$= \frac{B^2}{A} - \frac{2B^2}{A} + C =$
$\frac{AC - B^2}{A} < 0.$

This is clearly a contradiction because we have proved above that $f(\lambda) > 0$ for all real λ and hence $B^2 \not> AC$, therefore, $B^2 \leq AC$ or $B < A^{\frac{1}{2}} C^{\frac{1}{2}}$.

Now using the values of A, B and C, we get

$$\sum_{i=1}^{n} a_i b_i \leq (\sum_{i=1}^{n} a_i^2)^{\frac{1}{2}} (\sum_{i=1}^{n} b_i^2)^{\frac{1}{2}}.$$

Hence the result. $\square$

Definition 5.7. (Convex and Concave functions): Suppose a function $y = f(x)$ is defined in an interval I and is continuous there. If the chord joining any two points of the curve lies above the curve, then $f(x)$ is said to be convex downwards or simply convex.

The equation of chord joining end points $P[x_1, f(x_1)]$ and $Q[x_2, f(x_2)]$ is given by

$$y = f(x_1) + \tfrac{f(x_1)-f(x_2)}{x_1-x_2}(x - x_1)$$

Now $f(x)$ is convex, therefore, $f(x \leq f(x_1) + \tfrac{f(x_1)-f(x_2)}{x_1-x_2}(x - x_1)$, where x is a point on the curve lying between x_1 and x_2.

In particular, $f(\tfrac{x_1+x_2}{2}) \leq f(x_1) + \tfrac{f(x_1)-f(x_2)}{x_1-x_2}\tfrac{(x_2-x_1)}{2}$

or, $f(\tfrac{x_1+x_2}{2}) \leq f(x_1) - \tfrac{f(x_1)-f(x_2)}{x_1-x_2}\tfrac{(x_1-x_2)}{2}$

or, $f(\tfrac{x_1+x_2}{2}) \leq f(x_1) - \tfrac{f(x_1)-f(x_2)}{2}$

or, $f(\tfrac{x_1+x_2}{2}) \leq \tfrac{f(x_1)+f(x_2)}{2}$.

Therefore, $f(\tfrac{x_1+x_2}{2}) \leq \tfrac{f(x_1)+f(x_2)}{2}$.

This is also taken as the definition of a convex function $f(x)$.

Thus a function $f(x)$ is said to be convex in an interval I if

$$f(\tfrac{x_1+x_2}{2}) \leq \tfrac{f(x_1)+f(x_2)}{2}, \text{ where } x_1, x_2 \in I \text{ and } x_1 \neq x_2$$

If

$$f(\tfrac{x_1+x_2}{2}) \geq \tfrac{f(x_1)+f(x_2)}{2}, \text{ where } x_1, x_2 \in I \text{ and } x_1 \neq x_2,$$

then $f(x)$ is said to be concave in I.

Theorem 5.8. *Prove that if $x_1, x_2, ..., x_n \in I$ and f is continuous in I, then*

$$f\left(\frac{x_1+x_2+\dots+x_n}{n}\right) \leq \frac{f(x_1)+f(x_2)+\dots+f(x_n)}{n}$$

Proof. It is given that $x_1, x_2, \dots, x_n \in I$ and f is continuous in I.

Since f is convex in I and $x_1, x_2 \in I$, therefore,

$$\frac{f(x_1+x_2)}{2} \leq \frac{f(x_1)+f(x_2)}{2} \qquad (147)$$

This shows that result is true for $n = 2$.

Now suppose $n = 4$, and $x_1, x_2, x_3, x_4 \in I$. Then

$$f\left(\frac{x_1+x_2+x_3+x_4}{4}\right) = f\left(\frac{\frac{x_1+x_2}{2}+\frac{x_3+x_4}{2}}{2}\right)$$

$$\leq \frac{f\left(\frac{x_1+x_2}{2}\right)+f\left(\frac{x_3+x_4}{2}\right)}{2} \quad \text{(because } f \text{ is convex)}$$

$$\leq \frac{\frac{f(x_1)+f(x_2)}{2}+\frac{f(x_3)+f(x_4)}{2}}{2} \quad \text{by (147)}$$

$$= \frac{f(x_1)+f(x_2)+f(x_3)+f(x_4)}{4}.$$

Therefore, $\frac{f(x_1+x_2+x_3+x_4)}{4} \leq \frac{f(x_1)+f(x_2)+f(x_3)+f(x_4)}{4}$

This shows that the result is true for $n = 4 = 2^2$. Similarly result can be proved for $2^3, 2^4, \dots$. In fact the result can similarly be proved for any integral power of 2 say 2^m.

Now suppose n is a positive integer not of the form 2^m. We choose m in such a way that $2^m > n$. Now $2^m - n$ is a positive integer. Let $\frac{x_1+x_2+\dots+x_n}{n} = k$. Now consider 2^m numbers $x_1, x_2, \dots, x_n, k, k, k\dots k$. There are 2^{m-n} k's. Since the result has already been proved for positive integral powers of 2, therefore, result is true for 2^m also.

So from above, we get

$$f\left(\frac{x_1+x_2+\dots+x_n+k+k+\dots+k}{2^m}\right) = \frac{f(nk+(2^m-n)k)}{2^m} = f(k) = f\left(\frac{x_1+x_2+\dots+x_n}{n}\right)$$

Now $f\left(\frac{x_1+x_2+\dots+x_n+k+k+\dots+k}{2^m}\right) \leq \frac{f(x_1)+f(x_2)+\dots+f(x_n)+f(k)+f(k)+\dots+f(k)}{2^m}$

i.e, $f(k) \leq \frac{f(x_1)+f(x_2)+\dots+f(x_n)+(2^m-n)f(k)}{2^m}$

i.e, $2^m f(k) \leq f(x_1) + f(x_2) + ... + f(x_n) + (2^m)f(k) - (n)f(k)$

i.e, $nf(k) \leq f(x_1) + f(x_2) + ... + f(x_n)$

or $f(k) = \frac{f(x_1)+f(x_2)+...+f(x_n)}{n}$

i.e, $f(\frac{x_1+x_2+...+x_n}{n}) \leq \frac{f(x_1)+f(x_2)+...+f(x_n)}{n}$.

This shows that the result is true for the integers of the form 2^m and hence for all n. $\qquad\square$

Theorem 5.9. *Suppose $f''(x)$ exists in $[\alpha, \beta] = I$ and $f''(x) \geq 0$ in I. Then f is convex in I.*

Proof. It is given that $f''(x)$ exists in $[\alpha, \beta] = I$ and $f''(x) \geq 0$ in I.

Let $t_1, t_2 \in I$, where $t_1 < t_2$. Then, applying Taylor's Theorem (4.14), to $f(x)$ in $[t_1, \frac{t_1+t_2}{2}]$, we get

$f(t_1) = f(\frac{t_1+t_2}{2}) + (t_1 - \frac{t_1+t_2}{2})f'(\frac{t_1+t_2}{2}) + \frac{1}{2!}(t_1 - \frac{t_1+t_2}{2})^2 f''(c_1)$, for some $c_1, t_1 < c_1 < \frac{t_1+t_2}{2}$

or,

$$f(t_1) = f(\frac{t_1 + t_2}{2}) + (\frac{t_1 - t_2}{2})f'(\frac{t_1 + t_2}{2}) + \frac{1}{2!}(\frac{t_1 - t_2}{2})^2 f''(c_1). \quad (148)$$

Similarly for $[\frac{t_1+t_2}{2}, t_2]$

$$f(t_2) = f(\frac{t_1 + t_2}{2}) + (\frac{t_2 - t_1}{2})f'(\frac{t_1 + t_2}{2}) + \frac{1}{2!}(\frac{t_2 - t_1}{2})f''(c_2) \quad (149)$$

for some $c_2, \frac{t_1+t_2}{2} < c_2 < t_2$

Adding (148) and (149), we get

$f(t_1) + f(t_2) = 2f(\frac{t_1+t_2}{2}) + \frac{1}{2!}(\frac{t_1-t_2}{2})f''c_1 + \frac{1}{2!}(\frac{t_2-t_1}{2})f''(c_2)$, where $t_1 < c_1 < \frac{t_1+t_2}{2}; \frac{t_1+t_2}{2} < c_2 < t_2$

Now by hypothesis $f''(x) \geq 0$ in I and since $t_1, t_2 \in I$ and

$t_1 < c_1 < \frac{t_1+t_2}{2}; \frac{t_1+t_2}{2} < c_2 < t_2$, therefore, $f''(c_1) \geq 0$ and $f''(c_2) \geq 0$.

Therefore, $f(t_1) + f(t_2) \geq 2f(\frac{t_1+t_2}{2})$

or, $2f(\frac{t_1+t_2}{2}) \leq f(t_1) + f(t_2)$.

This gives $f(\frac{t_1+t_2}{2}) \leq \frac{f(t_1)+f(t_2)}{2}$.

This shows that f is convex in I. $\qquad\qquad\square$

Theorem 5.10. *Suppose $f''(x)$ exists in $[\alpha, \beta] = I$ and $f''(x) \leq 0$ in I. Then f is concave in I.*

Proof. It is given that $f''(x)$ exists in $[\alpha, \beta] = I$ and $f''(x) \leq 0$ in I. Let $t_1, t_2 \in I$, where $t_1 < t_2$, therefore, applying Taylor's theorem (4.14), to $f(x)$ in $[t_1, \frac{t_1+t_2}{2}]$, we get

$f(t_1) = f(\frac{t_1+t_2}{2}) + (t_1 - \frac{t_1+t_2}{2})f'(\frac{t_1+t_2}{2}) + \frac{1}{2!}(t_1 - \frac{t_1+t_2}{2})^2 f''(c_1)$, for some c_1, $t_1 < c_1 < \frac{t_1+t_2}{2}$

or,

$$f(t_1) = f(\frac{t_1 + t_2}{2}) + (\frac{t_1 - t_2}{2})f'(\frac{t_1 + t_2}{2}) + \frac{1}{2!}(\frac{t_1 - t_2}{2})^2 f''(c_1). \quad (150)$$

Similarly for $[\frac{t_1+t_2}{2}, t_2]$,

$$f(t_2) = f(\frac{t_1 + t_2}{2}) + (\frac{t_2 - t_1}{2})f'(\frac{t_1 + t_2}{2}) + \frac{1}{2!}(\frac{t_2 - t_1}{2})f''(c_2) \quad (151)$$

for some c_2, $\frac{t_1+t_2}{2} < c_2 < t_2$

Adding (150) and (151), we get

$f(t_1) + f(t_2) = 2f(\frac{t_1+t_2}{2}) + \frac{1}{2!}(\frac{t_1-t_2}{2})f''c_1 + \frac{1}{2!}(\frac{t_2-t_1}{2})f''(c_2)$,
$t_1 < c_1 < \frac{t_1+t_2}{2}; \frac{t_1+t_2}{2} < c_2 < t_2$

Now by hypothesis $f''(x) \leq 0$ in I and since $t_1, t_2 \in I$ and $t_1 < c_1 < \frac{t_1+t_2}{2}; \frac{t_1+t_2}{2} < c_2 < t_2$, therefore, $f''(c_1) \leq 0$ and $f''(c_2) \leq 0$.

So we have, $f(t_1) + f(t_2) \leq 2f(\frac{t_1+t_2}{2})$

or $2f(\frac{t_1+t_2}{2}) \geq f(t_1) + f(t_2)$.

This gives $f(\frac{t_1+t_2}{2}) \geq \frac{f(t_1)+f(t_2)}{2}$.

This shows that f is concave in I. $\square$

Remark 5.11. We know that if f is convex in I, then for $x_1, x_2 \in I$,

$$f(\tfrac{x_1+x_2}{2}) \leq \frac{f(x_1)+f(x_2)}{2}.$$

Taking $x_2 = x + h$ and $x_1 = x - h$ in above, we get

$$f(\tfrac{x-h+x+h}{2}) \leq \frac{f(x-h)+f(x+h)}{2}$$

or, $f(x) \leq \frac{f(x-h)+f(x+h)}{2}$

or, $2f(x) \leq f(x-h) + f(x+h) \Rightarrow f(x+h) + f(x-h) - 2f(x) \geq 0$.

Similarly if f is concave in I, then $f(x+h) + f(x-h) - 2f(x) \leq 0$.

Theorem 5.12. *Suppose f is convex in I and $f''(x)$ exists in I. Then $f''(x) \geq 0$ in I.*

Proof. Since f is given to be convex in I, therefore, for $x_1, x_2 \in I$

$$f(\tfrac{x_1+x_2}{2}) \leq \frac{f(x_1)+f(x_2)}{2}$$

Taking $x_2 = x + h$ and $x_1 = x - h$ in above we get

$$f(\tfrac{x-h+x+h}{2}) \leq \frac{f(x-h)+f(x+h)}{2}$$

or, $f(x) \leq \frac{f(x-h)+f(x+h)}{2}$.

Therefore,

$$f(x+h) + f(x-h) - 2f(x) \geq 0 \tag{152}$$

Now we know that if $f''(x)$ exists finitely ,then

$$f''(x) = lim_{h \to 0} \frac{f(x+h)+f(x-h)-2f(x)}{h^2}.$$

Therefore, from (152) it follows that $f''(x) \geq 0$. Hence the result follows. $\qquad\square$

Theorem 5.13. *Suppose f is concave in I and $f''(x)$ exists in I then $f''(x) \leq 0$ in I.*

Proof. Since f is given to be concave in I, therefore, for $x_1, x_2 \in I$

$$f(\tfrac{x_1+x_2}{2}) \geq \frac{f(x_1)+f(x_2)}{2}$$

Taking $x_2 = x + h$ and $x_1 = x - h$ in above we get

$$f(\tfrac{x-h+x+h}{2}) \geq \frac{f(x-h)+f(x+h)}{2}$$

or, $f(x) \leq \frac{f(x-h)+f(x+h)}{2}$

Therefore,

$$f(x + h) + f(x - h) - 2f(x) \leq 0 \tag{153}$$

Now we know that if $f''(x)$ exists finitely, then

$$f''(x) = lim_{h \to 0} \frac{f(x+h)+f(x-h)-2f(x)}{h^2}.$$

Therefore, from (153) it follows that $f''(x) \leq 0$.

Hence the result follows. $\qquad\square$

Theorem 5.14. (Jensen's Inequality): *Suppose f is convex in I and $f''(x)$ exists in I. Then if $t_1, t_2, ..., t_n$ are some points in I and $a_1, a_2, ..., a_n$ are positive numbers, then*

$$f(\tfrac{a_1 t_1+a_2 t_2+...+a_n t_n}{a_1+a_2+...+a_n}) \leq \frac{a_1 f(t_1)+a_2 f(t_2)+...+a_n f(t_n)}{a_1+a_2+...+a_n}$$

Proof. It is given that f is convex in I and $f''(x)$ exists in I, $t_1, t_2, ..., t_n$ are some points in I and $a_1, a_2, ..., a_n$ are the positive numbers.

Since f is convex in I and $f''(x)$ exists, therefore, $f''(x) \geq 0$ in I.

Let $\beta = \frac{a_1 t_1 + a_2 t_2 + ... + a_n t_n}{a_1 + a_2 + ... + a_n}$

i.e, $\beta = \frac{\sum_{r=1}^{n} a_r t_r}{\sum_{r=1}^{n} a_r}$

i.e, $\beta \sum_{r=1}^{n} a_r = \sum_{r=1}^{n} a_r t_r$

or,

$$\sum_{r=1}^{n} a_r t_r - \beta \sum_{r=1}^{n} a_r = 0 \tag{154}$$

Since $f''(x)$ exists in I, therefore, applying Taylor's Theorem (4.14), to $f(x)$, we get

$$f(t_r) = f(\beta) + (t_r - \beta)f'(\beta) + \tfrac{1}{2!}(t_r - \beta)^2 f''(c),$$

where $r = 1, 2, 3, ..., n$, and c lies between t_r and β

Multiplying this equation by a_r, we get

$a_r f(t_r) = a_r f(\beta) + (t_r - \beta)a_r f'(\beta) + \tfrac{1}{2!}(t_r - \beta)^2 a_r f''(c)$, where c lies between t_r and β

Adding these n equations, we get

$\sum_{r=1}^{n} a_r f(t_r)$
$= f(\beta) \sum_{r=1}^{n} a_r + f'(\beta)(\sum_{r=1}^{n} a_r t_r - \beta \sum_{r=1}^{n} a_r) + \tfrac{1}{2!} f''(c) \sum_{r=1}^{n} a_r (t_r - \beta)^2$

Now by using (154), we have,

$$\sum_{r=1}^{n} a_r f(t_r) = f(\beta) \sum_{r=1}^{n} a_r + P \tag{155}$$

where $P = \tfrac{1}{2!} f''(c) \sum_{r=1}^{n} a_r (t_r - \beta)^2$

Now $P \geq 0$ and $f''(x) \geq 0$ in I (because $(t_r - \beta)^2 \geq 0, a_r > 0$ for $r = 1, 2, 3, ..., n$)

Therefore, using in (155), it follows that

$$\sum_{r=1}^{n} a_r f(t_r) \geq f(\beta) \sum_{r=1}^{n} a_r$$

i.e. $f(\beta) \sum_{r=1}^{n} a_r \leq \sum_{r=1}^{n} a_r f(t_r)$

i.e, $f(\beta) \leq \frac{\sum_{r=1}^{n} a_r f(t_r)}{\sum_{r=1}^{n} a_r}$

i.e.

$$f\left(\frac{a_1 t_1 + a_2 t_2 + ... + a_n t_n}{a_1 + a_2 + ... + a_n}\right) \leq \frac{a_1 f(t_1) + a_2 f(t_2) + ... + a_n f(t_n)}{a_1 + a_2 + ... + a_n}$$

Thus the result follows. $\square$

Corollary 5.15. *If f is concave in I and $f''(x)$ exists, then the above inequality will get reversed.*

Exercise 5.16. Deduce the Generalized Arithmetical Geometrical mean inequality from Jensen's Inequality (5.14).

Solution: Consider $f(t) = -\log t$. Then $f'(t) = \frac{-1}{t}$ and $f''(t) = \frac{1}{t^2}$

Clearly $f''(t)$ exists for $t > 0$. Also for $t > 0, f''(t) > 0$.

Thus $f''(t)$ exists and $f''(t) > 0$ and, therefore, f is convex.

Therefore, by Jensen's inequality (5.14), it follows that

$$f\left(\frac{a_1 t_1 + a_2 t_2 + ... + a_n t_n}{a_1 + a_2 + ... + a_n}\right) \leq \frac{a_1 f(t_1) + a_2 f(t_2) + ... + a_n f(t_n)}{a_1 + a_2 + ... + a_n}$$

where $a_1, a_2, ..., a_n$ are positive numbers and $t_1 > 0, t_2 > 0, ..., t_n > 0$.

As $f(t) = -log t$, it follows from the above inequality that

$$-log\left(\frac{a_1 t_1 + a_2 t_2 + ... + a_n t_n}{a_1 + a_2 + ... + a_n}\right) \leq \frac{-a_1 log f(t_1) - a_2 log f(t_2) - ... - a_n log f(t_n)}{a_1 + a_2 + ... + a_n}$$

or, $$log\left(\frac{a_1 t_1 + a_2 t_2 + ... + a_n t_n}{a_1 + a_2 + ... + a_n}\right) \geq \frac{a_1 log f(t_1) + a_2 log f(t_2) + ... + a_n log f(t_n)}{a_1 + a_2 + ... + a_n}$$

i.e. $$log\left(\frac{a_1 t_1 + a_2 t_2 + ... + a_n t_n}{a_1 + a_2 + ... + a_n}\right) \geq log(t_1^{a_1} . t_2^{a_2} ... t_n^{a_n})^{\frac{1}{a_1 + a_2 + ... + a_n}}$$

Thus, we have, $$\frac{a_1 t_1 + a_2 t_2 + ... + a_n t_n}{a_1 + a_2 + ... + a_n} \geq (t_1^{a_1} . t_2^{a_2} ... t_n^{a_n})^{\frac{1}{a_1 + a_2 + ... + a_n}}$$

which is generalized Arithmetical Geometrical mean inequality.

Putting $a_1 = a_2 = ... = a_n = 1$, we get,

$$\frac{t_1 + t_2 + ... + t_n}{n} \geq (t_1 t_2 ... t_n)^{\frac{1}{n}}$$

i.e. $A \geq G$.

Theorem 5.17. *(**Holder's Inequality**)*: $\sum_{i=1}^{n} a_i b_i \leq \left(\sum_{i=1}^{n} a_i^p\right)^{\frac{1}{p}} \left(\sum_{i=1}^{n} b_i^q\right)^{\frac{1}{q}}$, *where* $p > 1, q > 1, \frac{1}{p} + \frac{1}{q} = 1$ *and* $a_i, b_i \geq 0$, *for all* $i = 1, 2, 3, ..., n$.

Proof. Suppose $a_i, b_i \geq 0$, for all $i = 1, 2, ..., n$. Let $p > 1, q > 1$, $\frac{1}{p} + \frac{1}{q} = 1$.

Consider $f(x) = x^q; x > 0, q > 1$. Then $f''(x) = q(q-1)x^{q-2} > 0$ for $x > 0$. Thus $f''(x)$ exists and is greater than zero and therefore, $f(x)$ is convex.

Now if $p_i, t_i > 0, i = 1, 2, ..., n$, then by Jensen's Inequality (5.14), we get

$$f\left(\frac{p_1 t_1 + p_2 t_2 + ... + p_n t_n}{p_1 + p_2 + ... + p_n}\right) \leq \frac{p_1 f(t_1) + p_2 f(t_2) + ... + p_n f(t_n)}{p_1 + p_2 + ... + p_n}$$

i.e. $$\left(\frac{p_1 t_1 + p_2 t_2 + ... + p_n t_n}{p_1 + p_2 + ... + p_n}\right)^q \leq \frac{p_1 t_1^q + p_2 t_2^q + ... + p_n t_n^q}{p_1 + p_2 + ... + p_n}$$

i.e. $$\frac{p_1 t_1 + p_2 t_2 + ... + p_n t_n}{p_1 + p_2 + ... + p_n} \leq \frac{(p_1 t_1^q + P_2 t_2^q + ... + p_n t_n^q)^{\frac{1}{q}}}{(p_1 + p_2 + ... + p_n)^{\frac{1}{q}}}$$

i.e. $\dfrac{\sum_{i=1}^{n} p_i t_i}{\sum_{i=1}^{n} p_i} \leq \dfrac{\left(\sum_{i=1}^{n} p_i t_i^q\right)^{\frac{1}{q}}}{\left(\sum_{i=1}^{n} p_i\right)^{\frac{1}{q}}}$

i.e $\sum_{i=1}^{n} p_i t_i \leq \dfrac{\left(\sum_{i=1}^{n} p_i\right)^1 \left(\sum_{i=1}^{n} p_i t_i^q\right)^{\frac{1}{q}}}{\left(\sum_{i=1}^{n} p_i\right)^{\frac{1}{q}}}$

i.e. $\sum_{i=1}^{n} p_i t_i \leq \left(\sum_{i=1}^{n} p_i\right)^{1-\frac{1}{q}} \left(\sum_{i=1}^{n} p_i t_i^q\right)^{\frac{1}{q}}$

or

$$\sum_{i=1}^{n} p_i t_i \leq \left(\sum_{i=1}^{n} p_i\right)^{\frac{1}{p}} \left(\sum_{i=1}^{n} p_i t_i^q\right)^{\frac{1}{q}} \tag{156}$$

Taking $p_i = a_i^p$ and $p_i t_i^q = b_i^q$, we have

$p_i t_i = a_i^p \left(\dfrac{b_i^q}{p_i}\right)^{\frac{1}{q}}$

i.e. $p_i t_i = a_i^p \dfrac{b_i}{p_i^{\frac{1}{q}}}$

i.e. $p_i t_i = a_i^p \dfrac{b_i}{p_i^{\frac{p}{q}}}$

i.e $p_i t_i = a_i^{p-\frac{p}{q}} b_i = a_i^{p\left(1-\frac{1}{q}\right)} b_i$

i.e $p_i t_i = a_i^{p.\left(\frac{1}{p}\right)} b_i$

i.e $p_i t_i = a_i b_i$.

Substituting value of $(p_i t_i)$ in (156) and using the fact that also $p_i = a_i^p$ and $p_i t_i^q = b_i^q$, we get

$\sum_{i=1}^{n} a_i b_i \leq \left(\sum_{i=1}^{n} a_i^p\right)^{\frac{1}{p}} \left(\sum_{i=1}^{n} b_i^q\right)^{\frac{1}{q}}.$ $\qquad\square$

Corollary 5.18. *Taking $p = q = 2$ in Holder's Inequality (5.17), we get*

$\sum_{i=1}^{n} a_i b_i \leq \left(\sum_{i=1}^{n} a_i^2\right)^{\frac{1}{2}} \left(\sum_{i=1}^{n} b_i^2\right)^{\frac{1}{2}}.$ *which is Cauchy Schwartz Inequality (5.6).*

Exercise 5.19. Let $a + b + c = 1, a > 0, b > 0, c > 0$, then prove that $(\frac{1}{a} - 1)(\frac{1}{b} - 1)(\frac{1}{c} - 1) \geq 8$ and equality holds when $a = b = c = \frac{1}{3}$.

Solution: Let $a > 0, b > 0, c > 0$ and $a + b + c = 1$. Then, by Arithmetic mean equality (5.16), it follows that

$$\sqrt{ab} \leq \frac{a+b}{2}$$

or

$$2\sqrt{ab} \leq a + b \tag{157}$$

and equality holds for a = b.

Since $a > 0, b > 0, c > 0$, it follows by the same reasoning that

$$2\sqrt{bc} \leq b + c \tag{158}$$

and

$$2\sqrt{ca} \leq c + a \tag{159}$$

Now using (157), (158) and (159), we get
$$(\frac{1}{a} - 1)(\frac{1}{b} - 1)(\frac{1}{c} - 1)$$
$$= \frac{(1-a)(1-b)(1-c)}{abc}$$
$$= \frac{(b+c)(c+a)(a+b)}{abc}$$
$$\geq 2\sqrt{bc}2\sqrt{ca}2\sqrt{ab}$$
$$= \frac{8abc}{abc} = 8.$$

Therefore, $(\frac{1}{a} - 1)(\frac{1}{b} - 1)(\frac{1}{c} - 1) \geq 8$ and equality holds for $a = b = c = \frac{1}{3}$.

Exercise 5.20. Prove that of all the rectangles with a given perimeter, a square has the largest area.

Solution: Let l be the length of rectangle and b be its breadth. Then perimeter of rectangle is equal to $2l + 2b$.

Let $2l + 2b = 2c$. Then
$$l + b = c. \tag{160}$$

Now the area of the rectangle is given by $A = l.b$

Applying Arithmetic Geometric mean inequality (5.16) to l and b, we get

$$\sqrt{lb} \leq \frac{l+b}{2} \tag{161}$$

i.e. $\sqrt{lb} \leq \frac{c}{2}$

i.e. $lb \leq \frac{c^2}{4}$

or

$$A \leq \frac{c^2}{4} \tag{162}$$

Since equality in (161) holds when $l = b$, it follows from (162) that equality holds when $l = b$ i.e. the maximum value of $A = \frac{c^2}{4}$ which is attained for $l = b = \left(\frac{c}{2}\right)$

i.e. A is maximum when this rectangle is a square. Hence the result follows.

Exercise 5.21. Prove that of all the triangles with a given perimeter , the equilateral triangle has the maximum area.

Solution: Let a, b, c be the sides of the triangle. Then its perimeter is given by $2S = a + b + c$.

If A is the area of this triangle, then by Hero's formula, we have

$$A = \sqrt{s(s-a)(s-b)(s-c)}$$

or $A^2 = s(s-a)(s-b)(s-c)$

Applying Arithmetic Geometric mean inequality (5.16) to numbers $s - a, s - b, s - c$, we get

$$\{(s-a)(s-b)(s-c)\}^{\frac{1}{3}} \leq \frac{(s-a) + (s-b) + (s-c)}{3} \tag{163}$$

and equality holds when $s - a = s - b = s - c$ i.e if $a = b = c$. Now from (163), we get

$$\{(s-a)(s-b)(s-c)\}^{\frac{1}{3}} \leq \frac{s+s+s-(a+b+c)}{3} \text{ and equality holds when}$$
$a = b = c$

i.e. $\{(s-a)(s-b)(s-c)\}^{\frac{1}{3}} \leq \frac{s+s+s-2s}{3}$ and equality holds when
$a = b = c$

i.e. $\{(s-a)(s-b)(s-c)\}^{\frac{1}{3}} \leq \frac{s}{3}$ and equality holds when $a = b = c$

i.e. $(s-a)(s-b)(s-c) \leq \frac{s^3}{27}$ and equality hold when $a = b = c$

i.e $s(s-a)(s-b)(s-c) \leq \frac{s^4}{27}$ and equality hold when $a = b = c$.

Therefore, $A^2 \leq \frac{s^4}{27}$ and equality hold when $a = b = c$

or $A \leq \frac{s^2}{3\sqrt{3}}$ and equality hold when $a = b = c$.

The maximum value of A is $\frac{s^2}{3\sqrt{3}}$ and this value is attained for $a = b = c$ i.e. when the triangle is equilateral triangle. Hence the result follows.

Theorem 5.22. *(Minkoski's inequality): Deduce Minkoski's inequality from Cauchy Schwartz inequality (5.6).*

Minkoski's inequality is

$$\sum_{i=1}^{n}\{(a_i+b_i)^2\}^{\frac{1}{2}} \leq (\sum_{i=1}^{n} a_i^2)^{\frac{1}{2}} + (\sum_{i=1}^{n} b_i^2)^{\frac{1}{2}};$$

where a_i and b_i are real number's.

Proof. By Cauchy Schwartz inequality (5.6), we have

$$\sum_{i=1}^{n} a_i b_i \leq (\sum_{i=1}^{n} a_i^2)^{\frac{1}{2}}(\sum_{i=1}^{n} b_i^2)^{\frac{1}{2}} \qquad (164)$$

Now, $\sum_{i=1}^{n}\{(a_i + b_i)^2\} = \sum_{i=1}^{n} a_i^2 + \sum_{i=1}^{n} b_i^2 + 2\sum_{i=1}^{n} a_i \sum_{i=1}^{n} b_i$

Therefore, (164) implies that $\sum_{i=1}^{n}\{(a_i+b_i)^2\} \le \sum_{i=1}^{n} a_i^2 + \sum_{i=1}^{n} b_i^2 + 2(\sum_{i=1}^{n} a_i^2)^{\frac{1}{2}}(\sum_{i=1}^{n} b_i^2)^{\frac{1}{2}}$

i.e. $\sum_{i=1}^{n}\{(a_i + b_i)^2\} \le \{(\sum_{i=1}^{n} a_i^2)^{\frac{1}{2}} + (\sum_{i=1}^{n} b_i^2)^{\frac{1}{2}}\}^2$

raising power $\frac{1}{2}$ on both sides we get

$$\sum_{i=1}^{n}\{(a_i + b_i)^2\}^{\frac{1}{2}} \le (\sum_{i=1}^{n} a_i^2)^{\frac{1}{2}} + (\sum_{i=1}^{n} b_i^2)^{\frac{1}{2}};$$

which is Minkoski's inequality. $\qquad\qquad\qquad\qquad\qquad\qquad\square$

Exercise 5.23. Let $p > 1, q > 1, \frac{1}{p} + \frac{1}{q} = 1$ and $u \ge 0, v \ge 0$. Then, Prove that

$$\frac{u^p}{p} + \frac{v^p}{q} \ge uv \text{ with equality if and only if } u^p = v^q$$

Solution Taking $n = 2$ in generalized Arithmetic Geometric mean inequality (5.16), we get

$$\frac{a_1 t_1 + a_2 t_2}{a_1 + a_2} \ge (t_1^{a_1} t_2^{a_2})^{\frac{1}{a_1 + a_2}}, a_1, a_2 > 0, t_1, t_2 > 0 \text{ with equality if } t_1 = t_2$$

or,

$$\frac{a_1}{a_1 + a_2} t_1 + \frac{a_2}{a_1 + a_2} t_2 \ge t_1^{\frac{a_1}{a_1 + a_2}} . t_2^{\frac{a_2}{a_1 + a_2}} \qquad (165)$$

with equality if and only if $t_1 = t_2$

Taking $\frac{a_1}{a_1 + a_2} = \frac{1}{p}, \frac{a_2}{a_1 + a_2} = \frac{1}{q}, t_1 = u^p$ and $t_2 = v^q$ (where $u \ge 0, v \ge 0$.

So $p > 1, q > 1, \frac{1}{p} + \frac{1}{q} = 1$.

Now (165) implies that, we get

$\frac{u^p}{p} + \frac{v^q}{q} \geq uv$ and equality holds if and only if $u^p = v^q$.

Exercise 5.24. Prove that if all numbers are positive and $\frac{x}{a} \neq \frac{y}{b}$, then

$$x\log(\tfrac{x}{a}) + y\log(\tfrac{y}{b}) > (x+y)\log(\tfrac{x+y}{a+b}).$$

Solution: It is given that all the numbers are positive.

Consider the function

$$f(t) = t\log(t) \tag{166}$$

. Then $f't = logt + 1$ and $f''(t) = \frac{1}{t}$ which exists for $t \neq 0$. Also $f''(t) > 0$ for $t > 0$. Thus $f''(t)$ exists for $t > 0$ and $f''(t) > 0$ for $t > 0$. So it follows that f(t) is convex for $t > 0$.

Therefore, by Jensen's inequality (5.14),

$$f(\tfrac{a_1t_1+a_2t_2}{a_1+a_2}) \leq \tfrac{a_1f(t_1)+a_2f(t_2)}{a_1+a_2} \text{ and equalities hold if and only if } t_1 = t_2$$

Now using(166), we have

$$\tfrac{a_1t_1+a_2t_2}{a_1+a_2}log(\tfrac{a_1t_1+a_2t_2}{a_1+a_2}) \leq \tfrac{a_1t_1log(t_1)+a_2t_2log(t_2)}{a_1+a_2} \text{ and equality holds if}$$
and only if $t_1 = t_2$

or, $(a_1t_1 + a_2t_2)log(\tfrac{a_1t_1+a_2t_2}{a_1+a_2}) \leq a_1t_1log(t_1) + a_2t_2log(t_2)$
and equality holds if and only if

$$t_1 = t_2. \tag{167}$$

Taking $a_1 = a, a_2 = b, t_1 = \tfrac{x}{a}, t_2 = \tfrac{y}{b}$ in (167), we get

$(x+y)log(\tfrac{x+y}{a+b}) \leq xlog(\tfrac{x}{a}) + ylog(\tfrac{y}{b})$ and equality holds if and only if $\tfrac{x}{a} = \tfrac{y}{b}$

Therefore, if $\tfrac{x}{a} \neq \tfrac{y}{b}$, then,

$$(x+y)log(\tfrac{x+y}{a+b}) < xlog\tfrac{x}{a} + ylog\tfrac{y}{b}$$

i.e. $xlog\tfrac{x}{a} + ylog\tfrac{y}{b} > (x+y)log(\tfrac{x+y}{a+b}).$

Exercise 5.25. Let $a_n > 0$ and $s_n = a_1 + a_2 + ... + a_n$. Prove that

$$(1 + a_1)(1 + a_2)...(1 + a_n) \leq 1 + s_n + \frac{s_n^2}{2!} + ... + \frac{s_n^n}{n!}.$$

Solution: Suppose $a_n > 0$ and $s_n = a_1 + a_2 + ... + a_n$.

Applying Arithmetic Geometric mean inequality (5.16) to

$(1 + a_1), (1 + a_2), ..., (1 + a_n)$, we get

$$\{(1 + a_1)(1 + a_2)...(1 + a_n)\}^{\frac{1}{n}} \leq \frac{(1+a_1)+(1+a_2)+...+(1+a_n)}{n}$$

i.e. $\{(1 + a_1)(1 + a_2)...(1 + a_n)\}^{\frac{1}{n}} \leq \frac{n+(a_1+a_2+...+a_n)}{n}$

i.e. $\{(1 + a_1)(1 + a_2)...(1 + a_n)\}^{\frac{1}{n}} \leq \frac{n+s_n}{n}$

i.e. $\{(1 + a_1)(1 + a_2)...(1 + a_n)\}^{\frac{1}{n}} \leq 1 + \frac{s_n}{n}$

i.e.
$$(1 + a_1)(1 + a_2)...(1 + a_n) \leq (1 + \frac{s_n}{n})^n \tag{168}$$

Now $(1+\frac{s_n}{n})^n = 1+n\frac{s_n}{n}+\frac{n(n-1)}{2!}(\frac{s_n}{n})^2+...+\frac{n(n-1)(n-2)...(n-(n-1))}{n!}(\frac{s_n}{n})^n$

i.e. $(1+\frac{s_n}{n})^n = 1+s_n+\frac{1}{2!}(1-\frac{1}{n})s_n^2+...+\frac{1}{n!}(1-\frac{1}{n})(1-\frac{2}{n})...(1-\frac{n-1}{n})s_n^n$

i.e. $(1 + \frac{s_n}{n})^n \leq 1 + s_n + \frac{s_n^2}{2!} + ... + \frac{s_n^n}{n!}$

and equality holds if $n = 1$.

using this in (168), we get

$$(1 + a_1)(1 + a_2)...(1 + a_n) \leq 1 + s_n + \frac{s_n^2}{2!} + ... + \frac{s_n^n}{n!}$$

Hence the result.

Exercise 5.26. Prove that if the sum of n positive integers is nc, then their product does not exceed c^n and sum of their squares lies between nc^2 and n^2c^2.

Solution: Let $a_1, a_2, ..., a_n$ be positive integers such that

$$a_1 + a_2 + ... + a_n = nc \tag{169}$$

Applying Arithmetic Geometric mean inequality (5.16) to $a_1, a_2, ..., a_n$, it follows that

$(a_1 a_2 ... a_n)^{\frac{1}{n}} \leq \frac{a_1 + a_2 + ... + a_n}{n}$

i.e. $a_1 a_2 ... a_n \leq \left(\frac{a_1 + a_2 + ... + a_n}{n}\right)^n$

i.e. $a_1 a_2 ... a_n \leq \left(\frac{nc}{n}\right)^n$

i.e. $a_1 a_2 ... a_n \leq c^n$

Hence the first result.

Now, $a_1, a_2, ..., a_n$ are positive integers, therefore,

$$a_1^2 + a_2^2 + ... + a_n^2 \leq (a_1 + a_2 + ... + a_n)^2$$

Therefore, using (169), we have,

$$a_1^2 + a_2^2 + ... + a_n^2 \leq n^2 c^2 \tag{170}$$

Also by Cauchy Schwartz inequality (5.6), we have

$\left(\sum_{i=1}^n a_i\right)^2 \leq \left(\sum_{i=1}^n 1.a_i\right)^2 \leq n \sum_{i=1}^n a_i^2$ (where $b_i = 1$).

Therefore,

$\left(\sum_{i=1}^n 1.a_i\right)^2 \leq n \sum_{i=1}^n a_i^2$

i.e. $(nc)^2 \leq n \sum_{i=1}^{n} a_i^2$

i.e. $n^2 c^2 \leq n \sum_{i=1}^{n} a_i^2$

or

$$nc^2 \leq \sum_{i=1}^{n} a_i^2 \tag{171}$$

Therefore, using (170) and (171), we get

$$nc^2 \leq a_1^2 + a_2^2 + \ldots + a_n^2 \leq n^2 c^2$$

Hence it follows that sum of squares of n given positive integers lies between nc^2 and $n^2 c^2$.

Exercise 5.27. Prove that if sum of two positive numbers is 4c and sum of their squares is $8c^2$, then none of them can exceed $(\sqrt{3}+1)c$.

Solution: Let $\alpha_1, \alpha_2, \alpha_3, \alpha_4$ be the four given positive numbers such that

$$\alpha_1 + \alpha_2 + \alpha_3 + \alpha_4 = 4c \tag{172}$$

and

$$\alpha_1^2 + \alpha_2^2 + \alpha_3^2 + \alpha_4^2 = 8c^2 \tag{173}$$

We have to prove that none of these can exceed $(\sqrt{3}+1)c$.

If possible suppose one of these numbers (say α_4) exceeds $(\sqrt{3}+1)c$, i.e. $\alpha_4 > (\sqrt{3}+1)c$

Let $\alpha_4 = (\sqrt{3}+1)c + \epsilon$, where $\epsilon > 0$

Then from (172), we get

$$\alpha_1 + \alpha_2 + \alpha_3 = 4c - \alpha_4 = 4c - (\sqrt{3}+1)c - \epsilon = 3c - \sqrt{3}c - \epsilon$$

i.e

$$\alpha_1 + \alpha_2 + \alpha_3 = c - \sqrt{3}c - \epsilon \tag{174}$$

Also from (173), we have

$$\begin{aligned}
&\alpha_1^2 + \alpha_2^2 + \alpha_3^2 \\
&= 8c^2 - \alpha_4^2 \\
&= 8c^2 - \{(\sqrt{3}+1)c + \epsilon\}^2 \\
&= 4c^2 - 2\sqrt{3}c^2 - 2c\epsilon - 2\sqrt{3}c\epsilon - \epsilon^2
\end{aligned}$$

i.e.

$$\alpha_1^2 + \alpha_2^2 + \alpha_3^2 = 4c^2 - 2\sqrt{3}c^2 - 2c\epsilon - 2\sqrt{3}c\epsilon - \epsilon^2 \tag{175}$$

By Cauchy Schwartz inequality (5.6), we have

$$\left(\sum_{i=1}^{3} \alpha_i\right)^2 \leq 3 \sum_{i=1}^{3} 3\alpha_i^2$$

Therefore, using (174) and (175), we have

$$(c - \sqrt{3}c - \epsilon)^2 \leq 4c^2 - 2\sqrt{3}c^2 - 2c\epsilon - 2\sqrt{3}c\epsilon - \epsilon^2$$

on simplification this gives $8\sqrt{3}c\epsilon \leq -4\epsilon^2$

or $2\sqrt{3}c \leq -\epsilon$.

Since L. H. S is positive and R. H. S is negative, therefore, we get a contradiction (because $c > 0, \epsilon > 0$).

Hence our Supposition that $\alpha_4 > (\sqrt{3}+1)c$ must be wrong.

Therefore, it follows that none of the numbers can exceed $(\sqrt{3}+1)c$.

EXERCISES

(1) Show that the function f is convex if and only if the function $-f$ is concave.[f need not to be differentiable and value of $-f$ at any point x is $-f(x)$].

(2) The function f is concave, but not necessarily differentiable. Find the values of constants a and b for which the function $af(x) + b$ is concave.

(3) Prove that for any positive numbers $a_1, a_2, ..., a_n$ satisfying $a_1 a_2 ... a_n = 1$ the following relation is true:

$$a_1 + a_2 + ... + a_n \geq n$$

Hence deduce that the Arithmetic mean of n non-negative real numbers is greater than or equal to their Geometric mean.

(4) Prove Cauchy's Schwarz inequality for complex numbers, i.e. prove that if $a_1, a_2, ..., a_n$ and $b_1, b_2, ..., b_n$; are complex numbers, then

$|\sum_{i=1}^{n} a_i \overline{b_i}| \leq (\sum_{i=1}^{n} |a_i|^2)(\sum_{i=1}^{n} |b_i|^2).$

(5) If $f''(x) > 0$ for all $x \in \mathbb{R}$, then prove that
$\frac{f(x_1+x_2)}{2} < \frac{f(x_1)+f(x_2)}{2}.$

(6) Show that if $a, b > 0$ and each $p, q > 1$ and $\frac{1}{p} + \frac{1}{q} = 1$, then

$$\frac{a^p}{p} + \frac{b^p}{q} \geq ab$$

where equality holds when $a^p = b^q$.

(7) For non negative real numbers $a_1, a_2, ..., a_n$ and $b_1, b_2, ..., b_n$ establish

  (a) Holders inequality, i.e.

    $\sum_{i=1}^{n} a_i b_i \leq \{\sum_{i=1}^{n} a_i^p\}^{\frac{1}{p}}\} + \{\sum_{i=1}^{n} b_i^q\}^{\frac{1}{q}}\}.$
    Observe the equality holds if and only if $a_i^p = k b_i^q$.

  (b) Minkowski's inequality, i.e.

    $\{\sum_{i=1}^{n} (a_i + b_i)^t\}^{\frac{1}{t}} \leq [\sum_{i=1}^{n} a_i^t]^{\frac{1}{t}} + [\sum_{i=1}^{n} b_i^t]^{\frac{1}{t}}.$
    Observe that the inequality occurs if and only if $a_i = k b_i$, $i = 1, 2, ..., n, k \in R, t > 1$,
    where p and q are positive numbers such that $\frac{1}{p} + \frac{1}{q} = 1$

(8) If $f''(x) > 0$ on [a, b], then prove Jensen's inequality:

$$f\left(\frac{x_1+x_2+...+x_n}{n}\right) \leq \frac{f(x_1)+f(x_2)+...+f(x_n)}{n}$$

Chapter 6

EXISTENCE OF INTEGRAL

The process of integration, as given in elementary texts on integral calculus, is generally introduced as the inverse of differentiation. The theory of integration seems to run on the tracks of summation.

To formulate independent theory of integration, the process of integration is defined on some broad-based arithmetical concepts which are due to the German Mathematician G.F.B Riemann (1826 - 1866).

The present chapter is, based on the definition of a Riemann integral which depends very explicitly on the order structure of real line. Accordingly we begin by discussing the integration of real bounded functions on intervals.

We begin with some definitions.

Definition 6.1. Let $[a, b]$ be a closed interval. By a partition of $[a, b]$ we mean a finite set P of points $x_0, x_1, x_2, ..., x_n$, where

$$a = x_0 \leq x_1 \leq x_2 \leq ... \leq x_n = b.$$

The partition P consists of $n+1$ points. Clearly any number of partitions of $[a, b]$ can be considered.

$[x_0, x_1], [x_1, x_2], ..., [x_{i-1}, x_i], ..., [x_{n-1}, x_n]$ are the subintervals of $[a, b]$.

We shall use the same symbol Δx_i to denote the ith subinterval $[x_{i-1}, x_i]$ as also its length $x_i - x_{i-1}$. Thus

$$\Delta x_i = x_i - x_{i-1}, (i = 1, 2, 3, ..., n)$$

Definition 6.2. Let $[a, b]$ be a closed interval and P be the partition of $[a, b]$. Let f be a bounded real function on $[a, b]$. Evidently f is bounded on each subinterval corresponding to each partition P. Let M_i, m_i be the bounds (supremum and infimum) of f in Δx_i. Consider the sums,

$$U(P, f) = \sum_{i=1}^{\infty} M_i \Delta x_i = M_1 \Delta x_1 + M_2 \Delta x_2 + ... + M_n \Delta x_n,$$

$$L(P, f) = \sum_{i=1}^{\infty} m_i \Delta x_i = m_1 \Delta x_1 + m_2 \Delta x_2 + ... + m_n \Delta x_n$$

known as the upper and lower sums of f corresponding to the partition P.

If M, m are bounds of f in $[a, b]$, then we have,

$$m \leq m_i \leq M_i \leq M$$

i.e. $m\Delta x_i \leq m_i \Delta x_i \leq M_i \Delta x_i \leq M \Delta x_i.$

Putting $i = 1, 2, 3, ..., n$ and adding all inequalities, we get

$$m(b - a) \leq L(P, f) \leq U(P, f) \leq M(b - a), b \geq a \qquad (176)$$

Now each partition gives rise to a pair of sums, the upper sum and the lower sum. By considering all partitions of $[a, b]$, we get a set U of upper sums and a set L of lower sums. The inequality (176), shows that both these sets are bounded and so each set has supremum and infimum. The infimum of set of upper sums is called the **Upper Integral**, and the supremum of the set of lower sums is called the **Lower Integral** of f over $[a, b]$. Thus

$$\overline{\int}_a^b f \, dx = inf.(U) \text{ or } \inf.U(P, f), \text{ for all partitions } P$$

$$\underline{\int_a^b} f \; dx = sup.(L) \text{ or sup. } L(P, f), \text{ for all partitions } P$$

These two integrals may or may not be equal.

Definition 6.3. (Darboux's condition of integrability): When two integrals $\overline{\int_a^b} f dx$ and $\underline{\int_a^b} f \; dx$ are equal, i.e.

$$\overline{\int_a^b} f \; dx = \underline{\int_a^b} f \; dx = \int_a^b f dx,$$

we say that f is Riemann integrable (or simply integrable) over $[a, b]$ and the common value of these integrals is called Riemann Integral (or simply Integral) of f over $[a, b]$.

The fact that f is integrable we express by writing $f \in R$.

Evidently from (176) above

$$m(b - a) \leq \int_a^b f dx \leq M(b - a), b \geq a \tag{177}$$

Remark 6.4.   (1) The statement that $\int_a^b f(x) \; dx$ exists, implies that the function f is bounded and integrable over [a,b].

(2) The upper and lower integrals are defined for every bounded function but they may not necessarily be equal for every bounded function. There exist functions for which these integrals are not equal; such functions are not integrable.

Example 6.5. The constant function $f(x) = k$ is integrable and

$$\int_a^b k \; dx = k(b - a)$$

For any partition P of interval $[a, b]$, we have
$$L(P, f) = k\Delta x_1 + k\Delta x_2 + ... + k\Delta x_n$$

or $L(P, f) = k(\Delta x_1 + \Delta x_2 + ... + \Delta x_n)$

or $L(P, f) = k(b - a)$.

This implies that $\underline{\int} k dx = sup L(P, f) = k(b - a)$

Also, $U(P, f) = k\Delta x_1 + k\Delta x_2 + ... + k\Delta x_n$

or $U(P, f) = k(\Delta x_1 + \Delta x_2 + ... + \Delta x_n)$

or $U(P, f) = k(b - a)$.

This implies that

$$\overline{\int} k dx = inf U(P, f) = k(b - a)$$

Thus, $\underline{\int} k\, dx = \overline{\int} k\, dx = k(b - a)$.

Therefore, the constant function k is integrable, and

$$\int_a^b = k(b - a)$$

Example 6.6. The function f defined by

$$f(x) = 0, \text{ when } x \text{ is rational}$$

$$f(x) = 1, \text{ when } x \text{ is irrational}$$

is not integrable on any interval.

Let us consider a partition P of an interval $[a, b]$.

Now $U(P, f) = \sum_{i=1}^{n} M_i \Delta x_i$

or $U(P, f) = 1.\Delta x_1 + 1.\Delta x_2 + ... + 1.\Delta x_n = (b - a)$

and $L(P, f) = \sum_{i=1}^{n} m_i \Delta x_i = 0.\Delta x_1 + 0.\Delta x_2 + ... + 0.\Delta x_n = 0$

Therefore, $\overline{\int}_a^b f\ dx = \inf U(P, f) = (b - a)$

and $\underline{\int}_a^b f\ dx = \sup L(P, f) = 0$

Thus, $\overline{\int} f\ dx \neq \underline{\int} f\ dx$

Hence function f is not integrable.

Example 6.7. The function $f(x) = x^2$ is integrable on any interval $[0, k]$.

Let us consider the partition P of $[0, k]$ obtained by dividing the interval into n equal parts. Thus,
$[0, \frac{k}{n}, \frac{2k}{n}, ..., \frac{nk}{n}]$ is a partition of P,

$[(i - 1)\frac{k}{n}]^2$ and $[i\frac{k}{n}]^2$ are the lower and upper bounds of the function in Δx_i, and the length of each such interval is $\frac{k}{n}$.

Therefore, $U(P, x^2) = \frac{k^3}{n^3}(1^2 + 2^2 + ... + n^2)$

or, $U(P, x^2) = \frac{k^3}{n^3} \cdot \frac{n}{6}(n + 1)(2n + 1)$

or, $U(P, x^2) = \frac{k^3}{6}(1 + \frac{1}{n})(2 + \frac{1}{n})$. Also $L(P, x^2) = \frac{k^3}{n^3}(0 + 1^2 + 2^2 + ... + (n - 1)^2)$
or $L(P, x^2) = \frac{k^3}{6}(1 - \frac{1}{n})(2 - \frac{1}{n})$.

Thus, $Inf(U(p, x^2)) = \frac{k^3}{3} = sup(L(p, x^2))$.

Hence the function is integrable and $\int_0^k x^2\ dx = \frac{k^3}{3}$.

Note 1: If f is bounded and integrable on $[b, a]$ for $a > b$, we define

$\int_a^b f\ dx = -\int_b^a f\ dx$. Also $\int_a^b f\ dx = 0$, when $a = b$.

(**)**Inequalities for Integrals:**

For a bounded integrable function f on $[a, b]$,

$$m(b-a) \leq \int_a^b f\,dx \leq M(b-a), \tag{178}$$

If $b < a$, then

$$m(a-b) \leq \int_b^a f\,dx \leq M(a-b)$$

i.e. $-m(a-b) \geq -\int_b^a f\,dx \geq -M(a-b)$

or,

$$m(b-a) \geq \int_a^b f\,dx \geq M(b-a) \tag{179}$$

From these inequalities, we deduce the following:

(1) If f is bounded and integrable on $[a, b]$, then there exists a number λ lying between the bounds of f such that

$$\int_a^b f\,dx = \lambda(b-a)$$

(2) If f is continuous and integrable on $[a, b]$, then there exists a number c between a and b such that

$$\int_a^b f\,dx = f(c)(b-a)$$

(3) If f is bounded and integrable on $[a, b]$ and k is number such that

$|f(x)| \leq k$, for all $x \in [a, b]$, then $\left|\int_a^b f\,dx\right| \leq k|b-a|$

Let M and m be the bounds of $f(x)$.

Now $|f(x)| \leq k$, for all $x \in [a, b]$.

Therefore, $-k \leq f(x) \leq k$

So, $-k \leq f(x) \leq M \leq k$.

Therefore, for $b \geq a$, we have

$$-k(b-a) \leq m(b-a) \leq \int_a^b f\,dx \leq M(b-a) \leq k(b-a)$$

i.e. $\left|\int_a^b f\,dx\right| \leq k(b-a)$. Now if $b < a$, then we have

$$\int_b^a f\,dx \leq k|a-b|$$

So we have

$$\Rightarrow \left| \int_a^b f\,dx \right| \le k|b - a|$$

The result is trivial for $a = b$.

(4) If f is bounded and integrable on $[a, b]$ and $f(x) \ge 0$, for all $x \in [a, b]$, then

$$\int_a^b f\,dx \ge 0 \text{ if } b \ge a$$
$$\int_a^b f\,dx \le 0 \text{ if } b \le a$$

Since $f(x) \ge 0$, for all $x \in [a, b]$, therefore, the lower bound $m \ge 0$

Result follows from inequalities (1) and (2)[(**)] above.

(5) If f and g are bounded and integrable on $[a, b]$, such that $f \ge g$, then

$$\int_a^b f\,dx \ge \int_a^b g\,dx \text{ when } b \ge a$$
$$\text{and } \int_a^b f\,dx \le \int_a^b g\,dx \text{ when } b \le a$$

Now $f \ge g$ implies that $f - g \ge 0$ for all $x \in [a, b]$.
Using deduction (4) [(**)], we have
$\int_a^b (f - g)\,dx \ge 0$ if $b \ge a$

or, $\int_a^b f\,dx \ge \int_a^b g\,dx$ if $b \ge a$.

Similarly $\int_a^b f\,dx \le \int_a^b g\,dx$ if $b \le a$

Refinement of Partitions:

Definition 6.8. For any partition P, the length of the largest subinterval is called the **norm** or **mesh** of the partition and is denoted as $\mu(P)$, (or simply by μ).

Therefore,

$$\mu(P) = max\,\Delta x_i, (1 \le i \le n)$$

A partition P^* is said to be a **refinement** of P if $P^* \supset P$, i.e. every point of P is point of P^*.

We also say that P^* refines P or that P^* is finer than P.

If P_1 and P_2 are two partitions, then we say that P^* is their **common refinement** if $P^* = P_1 \cup P_2$.

Theorem 6.9. *If P^* is a refinement of the partition P, then for a bounded function f,*

i) $L(P^*, f) \geq L(P, f)$

ii) $U(P^*, f) \leq U(P, f).$

Proof. i) Suppose P^* contains just one point more that P. Let this extra point be ξ, and suppose that this point is in Δx_i, i.e. $x_{i-1} < \xi < x_i$.

As the function is bounded over entire interval $[a, b]$, it is bounded in every subinterval Δx_i; $(i = 1, 2, 3, ..., n)$. Let w_1 and w_2 and m_i be the greatest lower bounds (glbs) of f in the intervals $[x_{i-1}, \xi]$; $[\xi, x_i]$; $[x_{i-1}, x_i]$ respectively.

Clearly $m_i \leq w_1, m_1 \leq w_2.$

Hence $L(P^*, f) - L(P, f)$

$$= w_1(\xi - x_{i-1}) + w_2(x_i - \xi) - m_i(x_i - x_{i-1})$$

$$= (w_1 - m_i)(\xi - x_{i-1}) + (w_2 - m_i)(x_i - \xi)$$

$$\geq 0.$$

If P^* contains p points more that P, we repeat the above reasoning p times and arrive at

$$L(P^*, f) \geq L(P, f)$$

Similarly we can prove that

$$U(P^*, f) \le U(P, f).$$

$\square$

Corollary 6.10. *If a refinement P^* of P contains p points more than P, and $|f(x)| \le k$, for all $x \in [a, b]$, then*

$$L(P, f) \le L(P^*, f) \le L(P, f) + 2pk\mu$$

$$U(P, f) \ge U(P^*, f) \ge U(P, f) - 2pk\mu.$$

Proof: *From the previous theorem (6.9) it follows that, when P^* contains one point more that P, we have*

$$L(P^*, f) - L(P, f) = (w_1 - m_1)(\xi - x_{i-1}) + (w_2 - m_i)(x_i - \xi).$$

Since $|f(x)| \le k$, for all $x \in [a, b]$, therefore,

$$-k \le m_i \le w_1 \le k, \ i.e. \ 0 \le w_1 - m_i \le 2k.$$

Similarly $0 \le w_2 - m_i \le 2k$.

Therefore, $L(P^*, f) - L(P, f) \le 2k(\xi - x_{i-1}) + 2k(x_i - \xi)$

or, $L(P^*, f) - L(P, f) = 2k\Delta x_i$

or, $L(P^*, f) - L(P, f) \le 2k\mu$, where μ is the norm of P.

Now supposing that each additional point is introduced one by one, by repeating the above reasoning p times, we get

$$L(P^*, f) \le L(P, f) + 2pk\mu$$

$$Also \ L(P, f) \le L(P^*, f).$$

Therefore, $L(P, f) \leq L(P^*, f) \leq L(P, f) + 2pk\mu.$

The other part can be proved similarly.

Exercise 6.11. If P^* is a refinement of P then

$$U(P^*, f) - L(P^*, f) \leq U(P, f) - L(P, f)$$

Solution: From corollary (6.10), we have

$$L(P, f) \leq L(P^*, f) \leq L(P, f) + 2pk\mu$$

$$\text{and } U(P, f) \geq U(P^*, f) \geq U(P, f) - 2pk\mu$$

i.e. $U(P^*, f) \leq U(P, f)$ and $L(P^*, f) \geq L(P, f)$
Hence $U(P^*, f) - L(P^*, f) \leq U(P, f) - L(P, f).$

Theorem 6.12. *For any two partitions P_1 and P_2*

$$L(P_1, f) \leq U(P_2, f)$$

i.e. no upper sum can ever be less than any lower sum.

Proof. Let P^* be the common refinement of P_1, P_2, so that

$$P^* = P_1 \bigcup P_2.$$

Using previous theorem (6.9), we get

$$L(P_1, f) \leq L(P^*, f) \leq U(P^*, f) \leq U(P_2, f) \tag{180}$$

Thus, $L(P_1, f) \leq U(P_2, f)$. $\qquad\qquad\qquad\qquad\qquad\qquad\qquad\qquad\qquad$ $\square$

Corollary 6.13. *For any bounded function f,*

$$\underline{\int} f \ dx \leq \overline{\int} f \ dx$$

Proof: *For any two partitions P_1 and P_2, let P^* be the common refinement. Then*

$$L(P_1, f) \leq L(P^*, f) \leq U(P^*, f) \leq U(P_2, f) \qquad (181)$$

By keeping P_2 fixed and taking the l.u.b over all partitions P_1, (181) implies

$$\underline{\int} f dx \leq U(P_2, f) \qquad (182)$$

Taking g.l.b over all P_2 in (182), we get

$$\underline{\int} f dx \leq \overline{\int} f dx.$$

Theorem 6.14. **(Darboux's Theorem)***: If f is a bounded function on $[a, b]$, then for every $\epsilon > 0$, there corresponds $\delta > 0$ such that*

(1) $U(P, f) < \overline{\int}_a^b f dx + \epsilon$

(2) $L(P, f) > \underline{\int}_a^b f dx - \epsilon$

for every partition P of $[a, b]$ with norm $\mu(P) < \delta$.

Proof. **1**: Since f is bounded, therefore there exists $k > 0$, such that

$$|f(x)| \leq k, \text{ for all } x \in [a, b]$$

Also since the upper integral is the infimum (g.l.b) of the set of upper sums, therefore, for every $\epsilon > 0$ there exists a partition $P_1\{x_0, x_1, x_2, ..., x_p\}$ of $[a, b]$ such that

$$U(P_1, f) < \overline{\int_a^b} f dx + \frac{\epsilon}{2}. \tag{183}$$

Now the partition P has $p - 1$ points besides $x_0(= a)$ and $x_p(= b)$.

Let δ be a positive number such that

$$2k(p - 1)\delta = \frac{\epsilon}{2} \tag{184}$$

Let P be any partition with norm $\mu(P) < \delta$.

Let, further, P^* be a refinement of P and P_1 i.e. $P^* = P \cup P_1$.

As P^* is refinement of P having at most $p - 1$ more points than P, therefore, corollary (6.13) implies that

$U(P, f) - 2k(p - 1)\delta \leq U(P^*, f)$

or, $U(P, f) - 2k(p - 1)\delta \leq U(P_1, f)$.

Now using (183), we have

$U(P, f) - 2k(p - 1)\delta < \overline{\int_a^b} f dx + \frac{\epsilon}{2}$

Therefore, using (184), we have

$U(P, f) < \overline{\int_a^b} f dx + \frac{\epsilon}{2} + \frac{\epsilon}{2} = \overline{\int_a^b} f dx + \epsilon,$

for any partition P with norm $\mu(P) < \delta$.

2: Proof of (2) is similar to that of (1). $\square$

EXERCISES

(1) Let $f : [a, b] \to \mathbb{R}$ be a bounded function. Then show that for any partition P of $[a, b]$ we have

$$m(b - a) \leq L(P, f) \leq U(P, f) \leq M(b - a).$$

(2) Show that for every $\epsilon > 0$, there exists $P \in P[a, b]$ such that

$$U[P, f] < \int_a^{-b} f\, dx + \epsilon.$$

(3) Show that for every $\epsilon > 0$, there exists $P' \in P[a, b]$ such that

$$L[P, f] > \int_{-a}^{b} f\, dx - \epsilon.$$

(4) If P^* is a refinement of P and $\mid f(x) \mid \leq k$ for all $x \in [a, b]$, then show that

$$L(P, f) \leq L(P^*, f) \leq L(P, f) + 2pk\delta$$

and

$$L(P^*, f) - L(P, f) \leq (M - m)\delta p$$

where $\parallel P \parallel = \delta$ and P^* has p additional points than P.

(5) If P^* is a refinement of P, then show that

$$W(P^*, f) \leq W(P, f).$$

(6) Show that for $P_1, P_2 \in P[a, b]$,

$$L(P_1, f) \leq U(P_2, f).$$

(7) If $f : [a, b] \to \mathbb{R}$ be a bounded function, then show that

$$\int_{-a}^{b} dx \leq \int_a^{-b} f\, dx.$$

(8) For $I = [0, 4]$, calculate the norms of the following partitions :

a) $P_1 = (0, 1, 2, 4)$

b) $P_2 = (0, 2, 3, 4)$.

(9) Let $f : [a, b] \to \mathbb{R}$ be a bounded function. If P_1 and P_2 are partitions of $[a, b]$, then show that $L(P, f) \le U(P, f)$.

(10) Let $f : [a, b] \to \mathbb{R}$ be a bounded function, then show that $L(f) \le U(f)$.

(11) Define upper and lower Darboux sums. If $P \in P[a, b]$, then show that $U(P)$ and $L(P)$ are bounded and $U(P) \ge L(P) \forall P \in P[a, b]$.

(12) If f is bounded on $[a, b]$ and $P_1, P_2 \in P[a, b]$, then prove that

$$P_1 \supset P_2 \Rightarrow U(P_1) - L(P_1) \le U(P_2) - L(P_2).$$

<h1 align="center">Chapter7</h1>

<h1 align="center">CONDITIONS OF INTEGRABILITY</h1>

We have stated earlier that a bounded function is said to be integrable when the upper and lower integrals are equal. We now formalize and give the necessary and sufficient conditions for the integrability of a function in two forms.

Theorem 7.1. (First form): *A necessary and sufficient condition for the integrability of a bounded function f is that to every $\epsilon > 0$, there corresponds $\delta > 0$ such that for every partition P of $[a, b]$ with norm $\mu(P) < \delta$,*

$$U(P, f) - L(P, f) < \epsilon.$$

Proof. Let a bounded function f be integrable, then

$$\underline{\int_a^b} f dx = \overline{\int_a^b} f dx = \int_a^b f dx.$$

Let ϵ be any positive number. By Darboux's Theorem (6.14) there exists $\delta > 0$ such that for every partition P with norm $\mu(P) < \delta$,

$$U(P, f) < \overline{\int_a^b} f dx + \frac{\epsilon}{2} = \int_a^b f dx + \frac{\epsilon}{2} \qquad (185)$$

$$L(P, f) > \underline{\int_a^b} f dx - \frac{\epsilon}{2} = \int_a^b f dx - \frac{\epsilon}{2} \qquad (186)$$

or,

$$-L(P,f) < -\int_a^b f\,dx + \frac{\epsilon}{2} \tag{187}$$

From (185) and (186), we get

$$U(P,f) - L(P,f) < \epsilon,$$

for every partition P with norm $\mu(P) < \delta$.

Conversely, let ϵ be any positive number. For any partition P with norm $\mu(P) < \delta$(depending on ϵ), we are given that

$$U(P,f) - L(P,f) < \epsilon.$$

Also for any partition P we know that,

$$L(P,f) \leq \underline{\int_a^b} f\,dx \leq \overline{\int_a^b} f\,dx \leq U(P,f)$$

$$\Rightarrow \overline{\int_a^b} f\,dx - \underline{\int_a^b} f\,dx \leq U(P,f) - L(P,f) < \epsilon.$$

Since ϵ is arbitrary positive number, therefore we see that a non-negative number is less that every positive number. So it must be equal to zero.

$$\text{Therefore, } \overline{\int_a^b} f\,dx - \underline{\int_a^b} f\,dx = 0,$$

$$\text{or, } \overline{\int_a^b} f\,dx = \underline{\int_a^b} f\,dx$$

Hence f is integrable. $\qquad\square$

Note 1: The Theorem can also be stated as follows:

A necessary and sufficient conditions for the integrability of a bounded function f is that

$$lim\{U(P,f) - L(P,f)\} = 0$$

when the norm $\mu(P)$ of the partition tends to 0.

Theorem 7.2. (**Second form**): *A bounded function f is integrable on $[a,b]$ if and only if for every $\epsilon > 0$, there exists a partition P such that*

$$U(P,f) - L(P,f) < \epsilon.$$

Proof. : Let the function f be integrable. Then

$$\underline{\int_a^b} f dx = \overline{\int_a^b} f dx = \int_a^b f dx.$$

Let ϵ be a positive number.

Since the upper and lower integrals are the infimum and supremum respectively of the upper and lower sums, therefore, there exist partitions P_1 and P_2 such that

$$U(P_1, f) < \overline{\int_a^b} f dx + \tfrac{\epsilon}{2} = \int_a^b f dx + \tfrac{\epsilon}{2}, \text{ and}$$
$$L(P_2, f) > \underline{\int_a^b} f dx - \tfrac{\epsilon}{2} = \int_a^b f dx - \tfrac{\epsilon}{2}.$$

Let P be the common refinement of P_1 and P_2.
Therefore, $U(P, f) \leq U(P_1, f) < \int_a^b f dx + \tfrac{\epsilon}{2} < L(P_2, f) + \epsilon \leq L(P, f) + \epsilon$

Thus there exists a partition P such that

$$U(P, f) - L(P, f) < \epsilon.$$

Conversely, let ϵ be any positive number and P be a partition for which

$$U(P, f) - L(P, f) < \epsilon.$$

For any partition P, we know that

$$L(P, f) \leq \underline{\int_a^b} f\,dx \leq \overline{\int_a^b} f\,dx \leq U(P, f).$$

Therefore,

$$\overline{\int_a^b} f\,dx - \underline{\int_a^b} f\,dx \leq U(P, f) - L(P, f) < \epsilon.$$

Now the non negative number, being less than every positive numbers ϵ, must be zero.

Therefore,

$$\overline{\int_a^b} f\,dx = \underline{\int_a^b} f\,dx.$$

Hence f is integrable. $\qquad\qquad\square$

Theorem 7.3. *A function f is integrable over $[a,\ b]$ if and only if there is a number I such that for any $\epsilon > 0$, there is a number λ such that for any $\epsilon > 0$, there exists a partition P of $[a,\ b]$ such that*

$$|U(P, f) - \lambda| < \epsilon \ and \ |\lambda - L(P, f)| < \epsilon.$$

Proof. Let f be integrable over $[a, b]$. Then for $\epsilon > 0$, there exists a partition P such that

$$|U(P, f) - L(P, f)| < \epsilon.$$

Now if λ is a number between $L(P, f)$ and $U(P, f)$, then

$$|U(P, f) - \lambda| < |U(P, f) - L(P, f)| < \epsilon$$
$$\text{and } |\lambda - L(P, f)| < |U(P, f) - L(P, f)| < \epsilon.$$

Hence the result.

Conversely for $\epsilon > 0$, there exists a partition P such that

$|U(P, f) - \lambda| < \frac{\epsilon}{2}$ and $|\lambda - L(P, f)| < \frac{\epsilon}{2}$.
Therefore, $|U(P, f) - L(P, f)| < |U(P, f) - \lambda| + |\lambda - L(P, f)| < \frac{\epsilon}{2} + \frac{\epsilon}{2} = \epsilon$

Hence f is integrable. $\qquad\qquad\qquad\qquad\qquad\qquad\qquad\qquad\square$

Theorem 7.4. *A function f is integrable over $[a, b]$ if and only if there is a number λ such that for any $\epsilon > 0$, there exists δ such that for all partitions P with $\mu(P) < \delta$,*

$$|U(P, f) - \lambda| < \epsilon \text{ and } |\lambda - L(P, f)| < \epsilon.$$

Proof. The proof is similar to the last theorem 7.3. $\qquad\qquad\qquad\square$

Integrability of the sum and difference of integrable functions:

Theorem 7.5. *If f_1 and f_2 are bounded and integrable functions on $[a, b]$ then $f = f_1 + f_2$ is also integrable on $[a, b]$, and*

$$\int_a^b f_1 \, dx + \int_a^b f_2 \, dx = \int_a^b f \, dx.$$

Proof. Clearly f is bounded on $[a, b]$.

Let $P = \{a = x_0, x_1, x_2, ..., x_n = b\}$ be any partitions of $[a, b]$ and $M_i', m_i'; M_i'', m_i''$ and M_i, m_i be the bounds of f_1, f_2 and f respectively in Δx_i.
Therefore,

$$m_i' + m_i'' \leq m_i \leq M_i' + M_i'' \qquad\qquad (188)$$

Multiplying by Δx_i and adding all these inequalities for $i = 1, 2, ..., n$; we get

$$L(P, f_1) + L(P, f_2) \le L(P, f) \le U(P, f_1) + U(P, f_2) \qquad (189)$$

Now let ϵ be any positive number.

Since f_1, f_2 are integrable, therefore, we can choose $\delta > 0$ such that, for any partitions P with norm $\mu(P) < \delta$, we have

$$U(P, f_1) - L(P, f_2) < \frac{1}{2}\epsilon \quad and \quad U(P, f_2) - L(P, f_2) < \frac{1}{2}\epsilon \qquad (190)$$

Thus for any partition P with norm $\mu(P) < \delta$, we have from (189) and (190)

$$U(P, f) - L(P, f)$$

$$\le U(P, f_1) + U(P, F_2) - L(P, f_1) - L(P, f_2)$$

$$< \tfrac{1}{2}\epsilon + \tfrac{1}{2}\epsilon = \epsilon$$

Thus the function f is integrable.

Now we proceed to prove the second part.

Since f_1 and f_2 are integrable and ϵ is any positive number, therefore, for all partitions P whose norm $\mu(P) < \delta$, we have

$$U(P, f_1) < \int_a^b f_1 dx + \frac{1}{2}\epsilon \quad and \quad U(P, f_2) < \int_a^b f_2 dx + \frac{1}{2}\epsilon \qquad (191)$$

Now (6.14) implies that

$$\int_a^b f \, dx \le U(P, f) \le U(P, f_1) + U(P, f_2)$$

So using (191), we have

$$\int_a^b f \, dx \; \text{¡} \; \int_a^b f_1 \, dx + \int_a^b f_2 \, dx + \epsilon$$

Since ϵ is arbitrary, we conclude that

$$\int_a^b f dx \le \int_a^b f_1 dx + \int_a^b f_2 dx \qquad (192)$$

Proceeding with $(-f_1), (-f_2)$ in place of f_1, f_2 respectively, we get

$$\int_a^b (-f)dx \leq \int_a^b (-f_1)dx + \int_a^b (-f_2)dx \quad or \quad \int_a^b f dx \geq \int_a^b f_1 dx + \int_a^b f_2 dx \tag{193}$$

Now (192) and (193) imply

$$\int_a^b f_1 dx + \int_a^b f_2 dx = \int_a^b f dx.$$

$\square$

Theorem 7.6. *If f_1, f_2 are two bounded and integrable functions on $[a,b]$, then $f = f_1 - f_2$ is also integrable on $[a, b]$ and*

$$\int_a^b f dx = \int_a^b f_1 dx - \int_a^b f_2 dx.$$

Proof. Let $f = f_1 + (-f_2)$, so that f is bounded on $[a, b]$.

Let $P = a = x_0, x_1, x_2, ..., x_n = b$ be any partition on $[a, b]$ and $M_i', m_i'; M_i'', m_i''$ and M_i, m_i be the bounds of f_1, f_2 and f respectively in Δx_i. Clearly the bounds of $(-f_2)$ are $-m_i''$ and $-M_i''$
Therefore,

$$m_i' - M_i'' \leq m_i \leq M_i \leq M_i' - m_i'' \tag{194}$$

multiplying by Δx_i and adding all these inequalities for $i = 1, 2, ..., n$, we get

$$L(P, f_1) - U(P, f_2) \leq L(P, f) \leq U(P, f) \leq U(P, f_1) - L(P, f_2) \tag{195}$$

Let $\epsilon > 0$ be a given number.

Since f_1, f_2 are integrable, therefore, there exists $\delta > 0$ such that for any partition P whose norm $\mu(P) < \delta$, is we have

$$U(P, f_1) - L(P, f_1) < \frac{1}{2}\epsilon \quad and \quad U(P, f_2) - L(P, f_2) < \frac{1}{2}\epsilon \tag{196}$$

Thus for any partition P with norm $\mu(P) < \delta$, (195) and (196) imply,

$$U(P, f) - L(P, f) \leq U(P, f_1) - L(P, f_2) - L(P, f_1) + U(P, f_2) < \tfrac{1}{2}\epsilon + \tfrac{1}{2}\epsilon = \epsilon$$

Thus the function f is integrable.

Let us now prove the second part.

Since f_1, f_2 are integrable and ϵ is any positive number, therefore, by Darboux's theorem (6.14), there exists $\delta > 0$ such that for all partitions P with norm $\mu(P) < \delta$, we have

$$U(P, f_1) < \int_a^b f_1 dx + \frac{1}{2}\epsilon \quad and \quad U(P, f_2) < \int_a^b f_2 dx - \frac{1}{2}\epsilon \qquad (197)$$

Therefore, $\int_a^b f dx \leq U(P, f) \leq U(P, f_1) - L(P, f_2) \leq \int_a^b f_1 dx - \int_a^b f_2 dx + \epsilon.$

Since ϵ is arbitrary, we conclude that,

$$\int_a^b f dx \leq \int_a^b f_1 dx - \int_a^b f_2 dx.$$

Proceeding with $(-f_1)$ and $(-f_2)$ in place of f_1 and f_2 respectively, we get

$$\int_a^b f dx \geq \int_a^b f_1 dx - \int_a^b f_2 dx.$$

or

$$\int_a^b f dx = \int_a^b f_1 dx - \int_a^b f_2 dx.$$

$\square$

Theorem 7.7.   *(1) If a bounded function f is integrable on $[a, b]$, then it is also integrable on $[a, c]$ and $[c, b]$, where c is a point of $[a, b]$.*

(2) Conversely, if f is bounded and integrable on $[a, c]$ and $[c, b]$, then it is also integrable on $[a, b]$.

(3) Also in either case $\int_a^b f\, dx = \int_a^c f\, dx + \int_c^b f\, dx$

Proof. 1: Since $f \in R$ over $[a, b]$, therefore for $\epsilon > 0$, there exists a Partition P such that

$$U(P, f) - L(P, f) < \epsilon$$

Let P^* be the refinement of P such that

$$P^* = P \cup \{c\}.$$

Therefore,

$$L(P, f) \leq L(P^*, f) \leq U(P^*, f) \leq U(P, f) \tag{198}$$

$$U(P^*, f) - L(P^*, f) \leq U(P, f) - L(P, f) < \epsilon \tag{199}$$

Let P_1 and P_2 denotes the sets of points of P^* between $[a, c]$ and $[c, b]$ respectively. Clearly P_1 and P_2 are partitions of $[a, c]$ and $[c, b]$ respectively. Also $P^* = P_1 \cup P_2$. Now

$$U(P^*, f) = U(P_1, f) + U(P_2, f) \tag{200}$$

and

$$L(P^*, f) = L(P_1, f) + L(P_2, f) \tag{201}$$

Hence

$$\{U(P_1, f) - L(P_1, f)\} + \{U(P_2, f) - L(P_2, f)\}$$

$$= U(P^*, f) - L(P^*, f)$$

$$< \epsilon.$$

Since (200) and (201) are non negative , it follows that partitions P_1, P_2 exist such that

$$U(P_1, f) - L(P_1, f) < \epsilon,$$
$$U(P_2, f) - L(P_2, f) < \epsilon.$$

This implies that f is integrable on $[a, c]$ and $[c, b]$.

2: Let $f \in R$ over $[a, c]$ and $[c, b]$.

Then for $\epsilon > 0$, we can find a partitions P_1 and P_2 of $[a, c]$ and $[c, b]$ respectively such that

$$U(P_1, f) - L(P_1, f) < \frac{\epsilon}{2} \tag{202}$$

$$U(P_2, f) - L(P_2, f) < \frac{\epsilon}{2} \tag{203}$$

Let $P^* = P_1 \cup P_2$.

Clearly P^* is a partition of $[a, b]$.
Therefore,

$$U(P^*, f) - L(P^*, f)$$

$$= U(P_1, f) + U(P_2, f) - L(P_1, f) - L(P_2, f)$$

$$\leq \frac{\epsilon}{2} + \frac{\epsilon}{2}$$

$$= \epsilon.$$

Thus for $\epsilon > 0$, there exists a partition P^* of $[a, b]$ such that

$$U(P^*, f) - L(P^*, f) < \epsilon$$

Therefore, $f \in R$ over $[a, b]$.

3: We know that for any functions f and g, if $h = f + g$, then

$$inf.h \geq inf.f + inf.g$$

Now for any partitions P_1, P_2 of $[a, c]$, $[c, b]$ respectively, if

$$P^* = P_1 \bigcup P_2, \text{ then}$$
$$U(P^*, f) = U(P_1, f) + U(P_2, f).$$

Therefore, on taking the infimum for all partitions, we get

$$\overline{\int_a^b} f dx \geq \overline{\int_a^c} f dx + \overline{\int_c^b} f dx$$

But since f is integrable on $[a, c]$, $[c, b]$ and $[a, b]$,therefore,

$$\int_a^b f dx \geq \int_a^c f dx + \int_c^b f dx \qquad (204)$$

Proceeding with $(-f)$ in place of f, we get

$$\int_a^b f dx \leq \int_a^c f dx + \int_c^b f dx \qquad (205)$$

Now (204) and (205) imply that

$$\int_a^b f dx = \int_a^c f dx + \int_c^b f dx.$$

$\square$

Integrability of the product, quotient and the modulus of integrable functions

Lemma 7.8. *The oscillation of a bounded function f on an interval $[a, b]$ is the supremum of the set,*

$$\{|f(x_1) - f(x_2)|; x_1, x_2 \in [a, b]\} \text{ of numbers.}$$

Proof: *Let M, m be the bounds of f on $[a, b]$.
Now*

$$m \leq f(x_1), f(x_2) \leq M, \text{ for all } x_1, x_2 \in [a, b].$$

Therefore,

$$|f(x_1) - f(x_2)| \leq M - m; \tag{206}$$

for all $x_1, x_2 \in [a, b]$

i.e. $M - m$ is an upper bound of the set in question.

*Let $\epsilon > 0$ be any given number.
Since M is supremum of f, therefore, there exists $x' \in [a, b]$ such that*

$$f(x') > M - \frac{1}{2}\epsilon \tag{207}$$

Similarly there exists $x'' \in [a, b]$ such that

$$f(x'') > m + \frac{1}{2}\epsilon \tag{208}$$

(207) and (208) imply that there exists $x', x'' \in [a, b]$ such that

$$f(x') - f(x'') > M - m - \epsilon$$

or

$$|f(x') - f(x'')| > M - m - \epsilon \tag{209}$$

(206) and (209) imply that

$M - m$ is an upper bound and no number less that $M - m$ can be an upper bound of set in question.

Hence $M - m = \sup.\{|f(x_1) - f(x_2)| : x_1, x_2 \in [a, b]\}$

Theorem 7.9. *If f_1 and f_2 are two bounded and integrable functions on $[a, b]$, then their product $f_1 f_2$ is also bounded and integrable on $[a, b]$.*

Proof. Since f_1 and f_2 are bounded, therefore, there exists k such that for all $x \in [a, b]$,

$$|f_1 x| \le k, |f_2(x)| \le k.$$

i.e. $|f_1(x) f_2(x)| \le k^2$, for all $x \in [a, b]$

which implies that $f_1 f_2$ is bounded.

Let $P = [a = x_0, x_1, x_2, ..., x_n = b]$ be any partition of $[a, b]$.

Let $M_i', m_i'; M_i'', m_i''$ and M_i, m_i be the bounds of f_1, f_2 and $f_1 f_2$ respectively in Δx_i.

We have for all $x_1, x_2 \in \Delta x_i$,

$$(f_1 f_2)(x_2) - (f_1 f_2)(x_1)$$

$$= f_1(x_2) f_2(x_2) - f_1(x_1) f_2(x_1)$$

$$= f_2(x_2)[f_1(x_2) - f_1(x_1)] + f_1(x_1)[f_2(x_2) - f_2(x_1)],$$

Therefore, $|(f_1 f_2)(x_2) - (f_1 f_2)(x_1)|$

$$\le |f_2(x_2)|.|f_1(x_2) - f_1(x_1)| + |f_1(x_1)|.|f_2(x_2) - f_2(x_1)|$$

$$\le k(M_i' - m_i') + k(M_i'' - m_i'')$$

So
$$M_i - m_i \le k(M_i' + m_i') + k(M_i'' - m_i'') \tag{210}$$

Now let $\epsilon > 0$ be a given number.

Since f_1, f_2 are integrable, therefore there exists $\delta > 0$ such that for any partition P with norm $\mu(P) < \delta$,

$$U(P, f_1) - L(P, f_1) < \tfrac{\epsilon}{2k}, U(P, f_2 - L(P, f_2) < \tfrac{\epsilon}{2k}.$$

Therefore, from (210), by multiplying by Δx_i, and adding all such inequalities, we have for any partition P with $\mu(P) < \delta$,

$$U(P, f_1 f_2) - L(P, f_1, f_2) \le k[U(P, f_1) - L(P, f_1)] + k[U(P, f_2) - L(P, f_2)]$$

$$\le k(\tfrac{\epsilon}{2k}) + k(\tfrac{\epsilon}{2k}) = \epsilon$$

Thus $f_1 f_2$ is integrable on $[a, b]$. $\square$

Theorem 7.10. *If f_1 and f_2 are two bounded and integrable functions on $[a,\ b]$ and there exists some $\lambda > 0$ such that $|f(x)| \ge \lambda$, for all $x \in [a, b]$, then $\frac{f_1}{f_2}$ is bounded and integrable on $[a,\ b]$.*

Proof. Since f_1 and f_2 are bounded, therefore there exists positive numbers k and λ such that

$$|f_1(x)| \le k, \lambda \le |f_2(x)| \le k, \text{ for all } x \in [a, b]$$

i.e. $(\frac{f_1}{f_2})(x) = \frac{f_1(x)}{f_2(x)} \le \frac{k}{\lambda}$, for all $x \in [a, b]$

Therefore, $\frac{f_1}{f_2}$ is bounded.

Let $P = [a = x_0, x_1, x_2, ..., x_n = b]$ be a partition of $[a, b]$.

Let $M_i', m_i'; M_i'', m_i''$ and M_i, m_i be the bounds of f_1, f_2 and $\frac{f_1}{f_2}$ respectively in Δx_i.

We have for all $x_1, x_2 \in \Delta x_i$.

$$\left|\left(\tfrac{f_1}{f_2}\right)(x_2) - \left(\tfrac{f_1}{f_2}\right)(x_1)\right| = \left|\tfrac{f_1(x_2)}{f_2(x_2)} - \tfrac{f_1(x_1)}{f_2(x_1)}\right| = \left|\tfrac{f_2(x_1)[f_1(x_2) - f_1(x_1)] - f_1(x_1)[f_2(x_2) - f_2(x_1)]}{f_2(x_2) f_2(x_1)}\right|$$

or, $\left|\left(\frac{f_1}{f_2}\right)(x_2) - \left(\frac{f_1}{f_2}\right)(x_1)\right| \le \frac{k}{\lambda^2}|f_1 x_2 - f_1(x_1)| + \frac{k}{\lambda^2}|f_2(x_2) - f_2(x_1)|$

or, $\left|\left(\frac{f_1}{f_2}\right)(x_2) - \left(\frac{f_1}{f_2}\right)(x_1)\right| \le \left(\frac{k}{\lambda^2}\right)(M_i' - m_i') + \left(\frac{k}{\lambda^2}\right)(M_i'' - m_i'')$

So,

$$\Rightarrow M_i - m_i \le \left(\frac{k}{\lambda^2}\right)(M_i' - m_i') + \left(\frac{k}{\lambda^2}\right)(M_i'' - m_i'') \qquad (211)$$

Now, let $\epsilon > 0$ be a given number.

Since f_1, f_2 are integrable, therefore there exists $\delta > 0$ such that for any partition P with norm $\mu(P) < \delta$,

$$U(P, f_1) - L(P, f_1) < \tfrac{\epsilon\lambda^2}{2k}, U(P, f_2 - L(P, f_2) < \tfrac{\epsilon\lambda^2}{2k}.$$

Therefore, from (211) for any partition P with $\mu(P) < \delta$, we have

$$U(P, \tfrac{f_1}{f_2}) - L(P, \tfrac{f_1}{f_2})$$

$$\leq (\tfrac{k}{\lambda^2})[U(P, f_1) - L(P, f_1)] + k/\lambda^2)[U(P, f_2) - L(P, f_2)]$$

i.e. $U(P, \tfrac{f_1}{f_2}) - L(P, \tfrac{f_1}{f_2})$

$$\leq (\tfrac{k}{\lambda^2})(\tfrac{\epsilon\lambda^2}{2k}) + (\tfrac{k}{\lambda^2})(\tfrac{\epsilon\lambda^2}{2k}) = \epsilon.$$

Thus $\tfrac{f_1}{f_2}$ is integrable on $[a, b]$. $\qquad\square$

Theorem 7.11. *If f is bounded and integrable on $[a,\ b]$, then $|f|$ is also bounded and integrable on $[a,\ b]$. Moreover*

$$\left| \int_a^b f dx \right| \leq \int_a^b |f|\ dx.$$

Proof. Since f is bounded, therefore there exists k such that

$$|f(x)| \leq k, \text{ for all } x \in [a, b].$$

Therefore, $|f|$ is bounded.

Again, since f is integrable, there exists a partition

$$P = \{a = x_0, x_1, x_2, ..., x_n = b\} \text{ of } [a, b] \text{ such that}$$

$$U(P, f) - L(P, f) < \epsilon.$$

Let M_i, m_i and M_i', m_i' be the bounds of f and $|f|$ in Δx_i.

We have for all $x_1, x_2 \in \Delta x_i$,

$$\left| |f|(x_2) - |f|(x_1) \right| = \left| |f(x_2)| - |f(x_1)| \right| \leq \left| f(x_2) - f(x_1) \right| \leq M_i - m_i$$

i.e. $M_i' - m_i' \leq M_i - m_i$.

This implies that for any partition P,

$$U(P, |f|) - L(P, |f|) \leq U(P, f) - L(P, f) < \epsilon.$$

Hence $|f|$ is integrable on $[a, b]$.

Since $f(x), -f(x) \leq |f(x)| \leq |f|(x)$, for all $x \in [a, b]$, therefore by (5) [inequalities for integrals [(**)], chapter 6], we have

$\int_a^b f dx \leq \int_a^b f dx$

and $- \int_a^b f dx = \int_a^b (-f) dx \leq \int_a^b |f| dx$

Thus, $\left| \int_a^b f dx \right| \leq \int_a^b |f| dx.$ $\square$

NOTE 2: The converse of the above theorem is not true.

Consider for example, the function

$$f(x) = 1, \text{ when } x \text{ is rational,}$$
$$f(x) = -1, \text{ when } x \text{ is irrational.}$$

Here $\overline{\int_a^b} f dx = b - a, \underline{\int_a^b} f dx = -(b - a)$

Therefore, f is not integrable.

But $|f(x)| = 1$, for all x,

so $\int_a^b |f| dx$ exists and is equal to $(b - a)$.

Thus $|f|$ is integrable while f is not.

Theorem 7.12. *If f is integrable on $[a, b]$, then f^2 is also integrable on $[a, b]$.*

Proof. Since f is bounded on $[a, b]$, therefore $|f|$ is also bounded on $[a, b]$. Thus there exists M such that for all $x \in [a, b]$,

$$|f(x)| \le M$$

Also since f is integrable, therefore $|f|$ is also integrable on $[a, b]$, and therefore for $\epsilon > 0$, there exists a partition P of $[a, b]$ such that

$$U(P, |f|) - L(P, |f|) < \tfrac{\epsilon}{2M}$$

Also since $|f^2(x)| = |f(x)|^2 \le M$, therefore f^2 is bounded.

If M_i, m_i are the bounds of $|f|$ and $M_i^{'}, m_i^{'}$ are those of f^2 in Δx_i, then $M_i^{'} = M_i^2$ and $m_i^{'} = m_i^2$.

Also

$$U(P, f^2) - L(P < f^2) = \sum_{i=1}^{n}(M_i^{'} - m_i^{'})\Delta x_i = \sum_{i=1}^{n}(M_i^2 - m_i^2)\Delta x_i$$

$$\text{i.e. } U(P, f^2) - L(P, f^2) = \sum_{i=1}^{n}(M_i - m_i)(M_i + m_i)\Delta x_i$$

$$\text{i.e. } U(P, f^2) - L(P, f^2) \le 2M\{\sum_{i=1}^{n}(M_i - m_i)\Delta x_i\}$$

$$\text{i.e. } U(P, f^2) - L(P, f^2) \le 2M\{U(P, |f|) - L(P, |f|)\}$$

$$\text{So we have } U(P, f^2) - L(P, f^2) < 2M\tfrac{\epsilon}{2M} = \epsilon$$

Thus, $f^2 \in R$ on $[a, b]$. $\qquad\qquad\square$

Corollary 7.13. *If f_1 and f_2 are both integrable on $[a, b]$, then $f_1 f_2$ is also integrable on $[a, b]$.*

Proof: *Since f_1 and f_2 are both integrable on $[a, b]$, therefore, f_1^2, f_2^2 and $(f_1 + f_2)^2$ are all integrable on $[a, b]$*

Now $f_1 f_2 = \frac{1}{2}\{(f_1 + f_2)^2 - f_1^2 - f_2^2\}$.

Therefore, $f_1 f_2$ is integrable on $[a, b]$.

The Integral as Limit of Sum (Riemann sums)

Definition 7.14. Riemann Sum: Let $f : [a, b] \to \mathbb{R}$ be a bounded function.

Let $P = [a = x_0, x_1, x_2, ..., x_n = b]$ be a partition of $[a, b]$. Corresponding to this partition P, let us choose the points $t_1, t_2, ..., t_n$ such that $x_{i-1} \leq t_i \leq x_i$ $(i = 1, 2, ..., n)$. Then the sum $\sum_{i=1}^{n} f(t_i)\Delta x_i$ is called the Riemann sum of f for the partition P and is denoted by $S(P, f)$, i.e.

$$S(P, f) = \sum_{i=1}^{n} f(t_i)\Delta x_i$$

We say that $S(P, f)$ converges to α as $\mu(P) \to 0$,

i.e.

$$lim_{\mu(P) \to 0} S(P, f) = \alpha$$

if for every $\epsilon > 0$ there exists $\delta > 0$ such that

$$|S(P, f) - \alpha| < \epsilon,$$

for every partition $P = [a = x_0, x_1, x_2, ..., x_n = b]$ with norm $\mu(P) < \delta$

and for every choice of points t_i in $[x_{i-1}, x_i]$.

Definition 7.15. (Second definition of integrability)

A function f is said to be integrable on $[a, b]$ if $lim S(P, f)$ exists as $\mu(P) \to 0$, and

$$lim_{\mu(P) \to 0} S(P, f) = \int_a^b f dx$$

Thus we have given two definitions of integrability (def.6.3 and def.7.15). Now f is Riemann integrable over $[a, b]$ in the first sense if it satisfies Darboux's properties and f is Riemann integrable in second sense if it satisfies the above condition.

Theorem 7.16. *A bounded function f is Riemann integrable over $[a, b]$ in first sense if and only if it is Riemann integrable in second sense.i.e; (def.6.3 and def.7.15 are equivalent).*

Proof. Let $f : [a, b] \to \mathbb{R}$ be bounded and integrable in first sense. Then

$$\underline{\int_a^b} f dx = \overline{\int_a^b} f dx = \int_a^b f dx$$

Let ϵ be any positive number.

Now by Darboux's Theorem (4.19), there exists $\delta > 0$ such that for every partition P with norm $\mu(P) < \delta$,

$$U(P, f) < \overline{\int_a^b} f dx + \epsilon = \int_a^b f dx + \epsilon \tag{212}$$

and

$$L(P, f) > \underline{\int_a^b} f dx - \epsilon \int_a^b f dx - \epsilon \tag{213}$$

If t_i is any point of Δx_i, then we have

$$L(P, f) \leq \sum_{i=1}^{n} f(t_i)\Delta x_i \leq U(P, f) \qquad (214)$$

From (212), (213), (214) we deduce that for any $\epsilon > 0$, there exists $\delta > 0$ such that for every partition with norm $\mu(P) < \delta$,

$$\int_a^b f\,dx - \epsilon < \sum_{i=1}^{n} f(t_i)\Delta x_i < \int_a^b f\,dx + \epsilon$$

$$\text{i.e. } \left|\sum_{i=1}^{n} f(t_i)\Delta x_i - \int_a^b f\,dx\right| < \epsilon$$

Thus the function is integrable in second sense.

Now let f be integrable according to the second definition, i.e.

$$lim_{\mu(P)\to 0} S(P, f) \text{ exists.}$$

In other words, for every number $\epsilon > 0$, there exists $\delta > 0$ such that for every partition $P = x_0, x_1, ..., x_n$ with norm $\mu(P) < \delta$ and for every choice of points t_i in Δx_i, there exists a number α such that

$$\left|\sum_{i=1}^{n} f(t_i)\Delta x_i - \alpha\right| < \epsilon.$$

First we will show that f is bounded.

Suppose if possible that f is not bounded.

Now in particular for $\epsilon = 1$, there exists a partition P such that for every choice of t_i in Δx_i

$$\left|\sum f(t_i)\Delta x_i - \alpha\right| < 1,$$

$$\text{i.e. } \left|\sum f(t_i)\Delta x_i\right| < |\alpha| + 1.$$

Now f is not bounded in $[a, b]$, it must also be so in at least one subinterval, say Δx_m.

Let us take $t_i = x_i$, when $i \neq m$, i.e. every t_i except t_m is fixed and accordingly every term of sum $\sum f(t_i)\Delta x_i$ except the term $f(t_m)\Delta x_m$ is also fixed. Since f is not bounded in Δx_m, we can choose a point t_m in Δx_m, such that

$$\left|\sum f(t_i)\Delta x_i\right| > |\alpha| + 1,$$

and thus we arrive at a contradiction.

Hence the function f is bounded on $[a, b]$.

Now let ϵ be any positive number. Then there exists $\delta > 0$ such that for any partitions P with $\mu(P) < \delta$, we have

$$\alpha - \epsilon/2 < S(P, f) < \alpha + \epsilon/2 \tag{215}$$

We choose one such P. If we let the points t_i range over the intervals Δx_i and take the *lub* and *glb* of members of $S(P, f)$ obtained in this way , then (215), yields

$$\alpha - \frac{\epsilon}{2} \leq L(P, f) \leq U(P, f) \leq \alpha + \frac{\epsilon}{2} \tag{216}$$

$$\text{i.e. } U(P, f) - L(P, f) < \epsilon$$

Also $L(P, f) \leq \underline{\int_a^b} f dx \leq \overline{\int_a^b} f dx \leq U(P, f)$

Therefore, $\overline{\int_a^b} f dx - \underline{\int_a^b} f dx < U(P, f) - L(P, f) < \epsilon,$

or a non-negative number is less than every positive number. This implies that

$$\overline{\int_a^b} f dx - \underline{\int_a^b} f dx = 0$$

or, $\overline{\int_a^b} f dx = \underline{\int_a^b} f dx$

Hence the function is integrable. $\qquad\qquad\square$

Example 7.17. Show that

$$\int_1^2 f\,dx = \tfrac{11}{2}, \text{ where } f(x) = 3x + 1.$$

The function is bounded over $[1, 2]$.

Let $P = \{1 = x_0, x_1, ..., x_n = 2\}$ be a partition which divides $[1, 2]$ into n equal subintervals, each of length $\frac{2-1}{n} = \frac{1}{n}$. Now

$$\mu(P) = \tfrac{1}{n} \to 0 \text{ as } n \to \infty,$$

$$x_i = 1 + \tfrac{i}{n}, i = 1, 2, ..., n;$$

$$\Delta x_i = \tfrac{1}{n}, i = 1, 2, ..., n;$$

$$\sum_{i=1}^{n} \Delta x_i = n.\tfrac{1}{n} = 1$$

Let $t_i = x_i, i = 1, 2, ..., n$. Then $S(P, f) = \sum_{i=1}^{n} f(t_i)\Delta x_i$

i.e. $S(P, f) = \sum_{i=1}^{n} f(x_i)\Delta x_i = \sum_{i=1}^{n}(3x_i + 1)\Delta x_i$

i.e. $S(P, f) = \sum_{i=1}^{n}\{3(1 + \tfrac{i}{n}) + 1\}\Delta x_i = 4\sum_{i=1}^{n} \Delta x_i + \tfrac{3}{n^2} \sum_{i=1}^{n} i$

i.e. $S(P, f) = 4 + \tfrac{3}{n^2}.\tfrac{n(n+1)}{2} = \tfrac{11}{2} + \tfrac{3}{2n}.$

Proceeding to limits when $\mu(P) \to 0.$, we get

$$lim_{\mu(P) \to 0} S(P, f) = \tfrac{11}{2}$$

Since the limit exists, the function is integrable and

$$\int_1^2 f\,dx = lim\, S(P, f) = \tfrac{11}{2}$$

Example 7.18. Evaluate $\int_{-1}^{1} f\,dx$, where $f(x) = |x|$.

The function f is bounded and continuous on $[-1, 1]$, and

$$f(x) = -x, x \leq 0,$$
$$f(x) = -x, x \geq 0.$$

Let $P = \{-1 = x_0, x_1, ..., x_n = 0 = y_0, y_1, ..., y_n = 1\}$ be a partition which divides $[-1, 1]$ into $2n$ equal subintervals, each of length $\frac{1}{n}$, Then

$\mu(P) = \frac{1}{n} \to 0$ as $n \to \infty$,

$x_i = -1 + \frac{i}{n}, i = 1, 2, ..., n;$

i.e. $y_i = \frac{i}{n}, i = 1, 2, ..., n$

$\Delta x_i = \frac{1}{n} = \Delta y_i, i = 1, 2, ..., n;$

$\sum_{i=1}^{n} \Delta x_i = n = \sum_{i=1}^{n} \Delta y_i$

Let $t_i \in \Delta x_i$ and $t_i' \in \Delta y_i$, and let $t_i = x_i, i = 1, 2, ..., n;\ t_i' = y_i, i = 1, 2, ..., n$

Then, $S(P, f) = \sum_{i=1}^{n} f(t_i)\Delta x_i + \sum_{i=1}^{n} f(t_i')\Delta y_i$

i.e. $S(P, f) = \sum_{i=1}^{n} (-x_i)\Delta x_i + \sum_{i=1}^{n} y_i\Delta y_i$

i.e. $S(P, f) = \sum_{i=1}^{n} (1 - \frac{1}{n})\Delta x_i + \sum_{i=1}^{n} \frac{i}{n}\Delta y_i$

i.e. $S(P, f) = \sum_{i=1}^{n} \Delta x_i - \frac{1}{n^2} \sum_{i=1}^{n} i + \frac{1}{n^2} \sum_{i=1}^{n} i = 1$

Therefore, $lim_{\mu(P) \to 0} S(P, f) = 1$, and since the limit exists, the function is integrable and

$$\int_{-1}^{1} |x|dx = limS(P, f) = 1$$

Theorem 7.19. *If f, f_1, f_2, where $f = f_1 \pm f_2$ are bounded and integrable on $[a, b]$, then*

$$\int_a^b f\,dx = \int_a^b f_1\,dx \pm \int_a^b f_2\,dx$$

Proof. Let ϵ be any positive number and $f = f_1 + f_2$.

Now it is given that f_1, f_2 are integrable, therefore there exists $\delta > 0$ such that for every partition $P = \{x_0, x_1, ..., x_n\}$ with norm $\mu(P) < \delta$ and for every choice of points t_i in Δx_i,

$$\left| \sum_i f_1(t_i)\Delta x_i - \int_a^b f_1\,dx \right| < \tfrac{1}{2}\epsilon,$$

i.e. $\left| \sum_i f_2(t_i)\Delta x_i - \int_a^b f_2\,dx \right| < \tfrac{1}{2}\epsilon$

i.e. $\left| \sum_i \{(f_1 + f_2)(t_i)\}\Delta x_i - \{\int_a^b f_1\,dx + \int_a^b f_2\,dx\} \right| < \epsilon$

Thus $\int_a^b f\,dx = \int_a^b (f_1 + f_2)\,dx = \int_a^b f_1\,dx + \int_a^b f_2\,dx$

The case of $f = f_1 - f_2$ may be discussed similarly. $\square$

Corollary 7.20. *If f_1 and f_2 are integrable over $[a, b]$ and c_1, c_2 are any two constants, then $c_1 f_1 + c_2 f_2 \in R$*

Proof: *Let $f = c_1 f_1 + c_2 f_2$.*

For any partition P, we can write

$$S(P, f) = \sum_i f(t_i)\Delta x_i = c_1 \sum_i f_1 \Delta x_i + c_2 \sum_i f_2 \Delta x_i$$

$$\text{i.e. } S(P, f) = c_1 S(P, f_1) + c_2 S(P, f_2)$$

Since f_1 and f_2 are integrable, therefore, for $\epsilon > 0$ we can choose $\delta > 0$ such that for all partitions P with $\mu(P) < \delta$, we have

$$|S(P, f_1) - \int_a^b f_1\,dx| < \epsilon$$

$$\text{and } |S(P, f_2) - \int_a^b f_2\,dx| < \epsilon$$

Therefore, $|S(P, f) - c_1 \int_a^b f_1 dx - c_2 \int_a^b f_2 dx| \leq c_1\epsilon + c_2\epsilon$

i.e. $lim_{\mu(P) \to 0} S(P, f)$ *exists and equals*

$$c_1 \int_a^b f_1 dx + c_2 \int_a^b f_2 dx.$$

Hence $(c_1 f_1 + c_2 f_2)$ *are integrable, and*

$$\int_a^b (c_1 f_1 + c_2 f_2) dx = c_1 \int_a^b f_1 dx + c_2 \int_a^b f_2 dx$$

Theorem 7.21. *If a function f is bounded and integrable on each of the intervals $[a, c]$, $[c, b]$ and $[a, b]$ where c is a point of $[a, b]$, then*

$$\int_a^b f dx = \int_a^c f dx + \int_c^b f dx$$

Proof. Let $\epsilon > 0$ be given.

As f is integrable on each of intervals $[a, c]$, $[c, b]$ and $[a, b]$, there exists $\delta > 0$ such that for every partition $P = \{a = x_0, x_1, ..., x_n = b\}$ containing the point c, with norm $\mu(P) < \delta$ and for every choice of points t_i in Δx_i,

$$|\sum_{[a,c]} f(t_i) \Delta x_i - \int_a^c f dx| < \tfrac{1}{3}\epsilon,$$

$$|\sum_{[c,b]} f(t_i) \Delta x_i - \int_c^b f dx| < \tfrac{1}{3}\epsilon, \text{ and}$$

$$|\sum_{[a,b]} f(t_i) \Delta x_i - \int_a^b f dx| < \tfrac{1}{3}\epsilon.$$

But $\sum_{[a,c]} f(t_i) \Delta x_i + \sum_{[c,b]} f(t_i) \Delta x_i = \sum_{[a,b]} f(t_i) \Delta x_i$

therefore we deduce that,

$$\left| \int_a^b f\,dx - \int_a^c f\,dx - \int_c^b f\,dx \right| < \epsilon$$

$$\text{i.e. } \int_a^b f\,dx = \int_a^c f\,dx + \int_c^b f\,dx$$

$\square$

Theorem 7.22. *A function f is integrable over $[a,b]$ if and only if for $\epsilon > 0$ there exists $\delta > 0$ such that if P, P' are any two partitions of $[a,b]$ such that $\mu(P) < \delta$ and $\mu(P') < \delta$, then*

$$|S(P,f) - S(P',f)| < \epsilon.$$

Proof. First let f be integrable over $[a,b]$, and

$$\int_a^b f\,dx = I$$

Then, for $\epsilon > 0$ there exists $\delta > 0$ such that for all partitions P, P' of $[a,b]$ such that $\mu(P) < \delta$, $\mu(P') < \delta$ and all positions of t_i in Δx_i,

$$|S(P,f) - l| < \frac{\epsilon}{2} \tag{217}$$

and

$$|S(P',f) - l| < \frac{\epsilon}{2} \tag{218}$$

i.e. $|S(P,f) - S(P',f)| < |S(P,f) - l| + |S(P',f) - l| < \frac{\epsilon}{2} + \frac{\epsilon}{2} = \epsilon.$

Conversely for $\epsilon > 0$ there exists δ_1 such that for any partition $P, P' \mu(P) < \delta, \mu(P') < \delta$, we have

$$|S(P,f) - S(P,f')| < \frac{\epsilon}{2} \tag{219}$$

We know that for a given partition P' and every choice of t_i in Δx_i, $S(P',f)$ is bounded by $L(P',f)$ and $U(P',f)$ which for all partitions of $[a,b]$ are, in turn, bounded by $m(b-a)$ and $M(b-a)$, where m and M are bounds of f.

Thus the sequence $\{S(P',f)\}$ of Riemann sums is bounded. As every bounded sequence has a limit point, let the sequence has a limit point l. Therefore,

$$lim_{\mu(P')\to 0}S(P',f) = l$$

Therefore, for $\epsilon > 0$, there exists δ_2 such that for partition P' with $\mu(P') < \delta_2$,

$$|S(P',f) - l| < \frac{\epsilon}{2} \tag{220}$$

Let $\delta = min.(\delta_1, \delta_2)$. Then, for $\mu(P) < \delta$,

$$|S(P,f) - l| \le |S(P,f) - S(P',f)| + |S(P',f) - l| < \frac{\epsilon}{2} + \frac{\epsilon}{2} = \epsilon$$

$$i.e. lim_{\mu(P)\to 0}S(P,f) = l$$

Thus $lim_{\mu(P)\to 0}S(P,f)$ exists and hence the function f is integrable. $\quad\square$

Some Integrable Functions

Theorem 7.23. *Every continuous function is integrable.*

Proof. Let f be a function continuous on $[a,b]$. We will prove that f is also Riemann integrable on $[a,b]$.

Let $\epsilon > 0$ be given. Choose a positive number η, such that

$$\eta(b - a) < \epsilon$$

Now f is continuous on closed interval $[a,b]$, therefore, it is bounded and is uniformly continuous on $[a,b]$, which implies that there exists $\delta > 0$, such that

$$|f(t_1) - f(t_2)| < \eta, \tag{221}$$

if $|t_1 - t_2| < \delta$ and $t_1, t_2 \in [a,b]$.

We now choose a partition P with norm $\mu(p) < \delta$.

Then by (221) we have

$$M_i - m_i \leq \eta \quad (i = 1, 2, ..., n)$$

Hence,

$$U(P, f) - L(P, f) = \sum_{i=1}^{n}(M_i - m_i)\Delta x_i \leq \eta \sum_{i=1}^{n} \Delta x_i$$

$$\Rightarrow U(P, f) - L(P, f) \leq \eta(b - a) < \epsilon \qquad (222)$$

Thus f is integrable. $\qquad\qquad\square$

Corollary 7.24. *If a function f is continuous, then to every $\epsilon > 0$ there corresponds $\delta > 0$ such that*

$$|\sum_{i=1}^{n} f(t_i)\Delta x_i - \int_a^b fdx| < \epsilon$$

for every partition $P = [x_0, x_1, x_2, ..., x_n]$ of $[a, b]$ with $\mu(P) < \delta$, and for every choice of points t_i in $[x_{i-1}, x_i]$.

The proof follows from (222) above, Since the two numbers $\sum f(t_i)\Delta x_i$ and $\int fdx$ lie between $U(P, f)$ and $L(P, f)$.

Theorem 7.25. *If a function f is monotonic on $[a, b]$, then it is integrable on $[a, b]$.*

Proof. We will prove the Theorem when f is monotonic increasing (proof for other case is analogues).

Now f is bounded. Let $\epsilon > 0$ be given.

Let us choose a number $\eta < \frac{\epsilon}{f(b)-f(a)+1}$. We now choose a partition $P = \{x_0, x_1, x_2, ..., x_n\}$ of $[a, b]$ with norm $\mu(P) < \eta$.

Since f is monotonic increasing, therefore

$$M_i = f(x_i), m_i = f(x_{i-1}), (i = 1, 2, ..., n)$$

Hence

$$U(P, f) - L(P, f) = \sum_{i=1}^{n}(M_i - m_i)\Delta x_i$$

i.e. $U(P, f) - L(P, f) = \sum_{i=1}^{n}\{f(x_i) - f(x_{i-1})\}\Delta x_i$

i.e. $U(P, f) - L(P, f) < \eta \sum_{i=1}^{n}\{f(x_i) - f(x_{i-1})\}$

i.e. $U(P, f) - L(P, f) < \frac{\epsilon}{f(b)-f(a)+1} \cdot \{f(b) - f(a)\}$

or $U(P, f) - L(P, f) < \epsilon$

Hence f is integrable. $\qquad\qquad\square$

Theorem 7.26. *A bounded function f, having a finite number of points of discontinuity on $[a, b]$ is integrable on $[a, b]$.*

Proof. Let M and m be bounds of f and $\epsilon > 0$) be a given number.

Let there be p points of discontinuity of f on $[a, b]$.

We now consider a partition P of $[a, b]$ such that all the points of discontinuity get enclosed in p non-overlapping sub-intervals, the sum of whose lengths $< \frac{\epsilon}{2(M-m)}$. Now the oscillation of f in each of these sub-intervals is less than $(M - m)$, therefore, the total contribution to the difference $\{U(P, f) - L(P, f)\}$ is less than $\frac{\epsilon}{2(M-m)}(M - m) = \frac{\epsilon}{2}$.

Now the function f is continuous in remaining portion of $[a, b]$ i.e. in $(p + 1)$ sub-intervals of $[a, b]$, excluding the sub-intervals considered above.

As in theorem (7.23), the contribution to the difference $\{U(P, f) - L(P, f)\}$ from each of these $(p + 1)$ sub-intervals can be made $< \frac{\epsilon}{2(p+1)}$, and so the total contribution to $\{U(P, f) - L(P, f)\}$ by these $(p + 1)$ sub-intervals is less than $\frac{\epsilon}{2(p+1)}(p + 1) = \frac{\epsilon}{2}$.

Thus for partition P of $[a, b]$,

$$U(P, f) - L(P, f) < \epsilon.$$

Hence the function f is integrable.   □

Theorem 7.27. *A bounded function f is integrable on $[a, b]$, if the set of points of discontinuity has only a finite number of limit points.*

Proof. Let the set of points of discontinuity of f have a finite number p of limit points. Let M, m be the bounds of f.

The limit points may be enclosed in p non-overlapping subintervals of $[a, b]$, the sum of whose lengths is less than $\frac{\epsilon}{2(M-m)}$, such that their total contribution to $\{U(P, f) - L(P, f)\}$ is less than $\frac{\epsilon}{2}$.

Only a finite number of points of discontinuity of f can be outside these sub-intervals, i.e. the function f has a finite number of points of discontinuity on $[a, b]$ excluding the p sub-intervals enclosing the limit points. Therefore, as in theorem (7.26), the total contribution to $\{U(P, f) - L(P, f)\}$ from these sub-intervals of $[a, b]$ can be made $< \frac{\epsilon}{2}$.

Thus for such a partition P of $[a, b]$,

$$U(P, f) - L(P, f) < \epsilon.$$

Hence the function f is integrable on $[a, b]$.   □

Exercise 7.28. A function f is defined on $[0, 1]$ as follows:

$f(x) = 0$, when x is irrational or zero and

$f(x) = \frac{1}{q}$, when x is any non zero rational number $\frac{p}{q}$ in lowest form.

Show that f is integrable on $[0, 1]$ and the value of integral is zero.

Solution: The function f is bounded with bounds 0 and 1.

Let ϵ be any positive number. Then there exists a largest integer $q \in \mathbb{N}$ such that $\frac{1}{q} > \frac{\epsilon}{2}$ or $q < \frac{2}{\epsilon}$. Therefore, there are only a finite number of points $\frac{p}{q}$ for which $\frac{1}{q} > \frac{\epsilon}{2}$. Let us call such points as exceptional points.

Thus at those rational points which are exceptional points, f has value $\frac{1}{q} > \frac{\epsilon}{2}$, while at the other rational points , the value of f is $\frac{1}{q} < \frac{\epsilon}{2}$. The function is zero at irrational points.

Also, every interval contains rational as well as irrational points.

Thus the oscillation of f in any interval which includes no exceptional points is less than $\frac{\epsilon}{2}$ and that in an interval which includes the exceptional points it is at the most equal to 1.

Let us consider a partition P of $[0, 1]$ so as to enclose the exceptional points (finite in number) in sub-intervals, the sum of whose lengths is less than $\frac{\epsilon}{2}$. Thus the contribution to $\{U(P, f) - L(P, f)\}$ made by these is less than $\frac{\epsilon}{2}$.

The contribution to $\{U(P, f) - L(P, f)\}$ by the remaining portion of $[0, 1]$ is evidently less than $\frac{\epsilon}{2}$.

Hence

$$\{U(P, f) - L(P, f)\} < \epsilon$$

so that function f is integrable.

Exercise 7.29. Show that the function f defined as follows:

$$f(x) = \frac{1}{2^n}, \text{ when } \frac{1}{2^{n+1}} < x \leq \frac{1}{2^n}, \ (n = 1, 2, ...,);$$

$$f(0) = 0,$$

is integrable on $[0, 1]$, although it has an infinite number of points of discontinuity.

Solution: We have

$$f(x) = \frac{1}{2^n}, \text{ when } \frac{1}{2^{n+1}} < x \leq \frac{1}{2^n}, \ (n = 1, 2, ...,);$$

$$f(0) = 0,$$

Now

$$f(x) = 1, \text{ when } \tfrac{1}{2} < x \le \tfrac{1}{1}$$

$$f(x) = \tfrac{1}{2}, \text{ when } \tfrac{1}{2^2} < x \le \tfrac{1}{2}$$

$$f(x) = \tfrac{1}{2^2}, \text{ when } \tfrac{1}{2^3} < x \le \tfrac{1}{2^2}$$

$$\cdots \ \cdots \ \cdots \ \cdots \ \cdots$$

$$\cdots \ \cdots \ \cdots \ \cdots \ \cdots$$

$$f(x) = \tfrac{1}{2^{n-1}}, \text{ when } \tfrac{1}{2^n} < x \le \tfrac{1}{2^{n-1}}$$

$$\cdots \ \cdots \ \cdots \ \cdots \ \cdots$$

$$f(x) = 0, \text{ when } x = 0.$$

Thus we notice that f is bounded and monotonic increasing on $[0, 1]$.

Hence f is integrable.(Theorem 7.25)

EXERCISES

(1) Let $f, g : [a, b] \to \mathbb{R}$ be integrable functions. Then $f + g$ is also integrable and

$$\int_a^b (f + g)(x)dx \leq \int_a^b f(x)dx + \int_a^b g(x)dx.$$

(2) If $f(x) \leq g(x)$ for all $x \in [a, b]$, then show that

$$\int_a^b f = \int_a^b g.$$

(3) Suppose that $f : [a, b] \to \mathbb{R}$ and that $f(x) = 0$ except for a finite number of points $c_1, - - -, c_n$ in $[a, b]$. Prove that $f \in R[a, b]$ and that $\int_a^b f = 0$.

(4) Let $f, g : [a, b] \to \mathbb{R}$ be integrable, then show that $f.g$ is also integrable.If $f \leq g$, then

$$\int_a^b f(x)dx \leq \int_a^b g(x)dx.$$

(5) If f, g are integrable, then show that the function $\mid f \mid$ is integrable and

$$\mid \int_a^b f(x)dx \mid \leq \int_a^b \mid f \mid (x)dx.$$

(6) Let $f : [a, b] \to \mathbb{R}$. If f is continuous, then show that it is integrable.

(7) Let $f : [a, b] \to \mathbb{R}$. If f is monotonic, then show that it is integrable.

(8) Let $f : [0, 1] \to \mathbb{R}$ be given by

$$f(x) = \{1 + x \text{ if } x \text{ is rational,}$$

$$= 0 \text{ if } x \text{ is irrational.}$$

Is f integrable?

(9) If $f, g \in R[a, b]$, then prove that

$$\int_a^b f^2 dx + \int_a^b g^2 dx \leq 2 \int_a^b |fg| \, dx$$

(10) If $f \in R[a, b]$ and $|f(x)| \leq M$ for all $x \in [a, b]$, show that

$$\left| \int_a^b f \right| \leq M(b - a).$$

(11) If $f \in R[a, b]$, then prove that the value of the integral is uniquely determined.

(12) Suppose that f and g are in $R[a, b]$. Then if $k \in \mathbb{R}$, the function kf is in $R[a, b]$ and
$$\int_a^b kf = k \int_a^b f.$$

(13) Let $f(x) = 2$ if $0 \leq x < 1$ and $f(x) = 1$ if $1 \leq x \leq 2$. Show that $f \in R[0, 2]$.

(14) If f is bounded on (a, b) and is Riemann integral on every closed subinterval of (a, b), then show that f is integrable on (a, b).

Chapter8

FUNDAMENTAL THEOREMS

We shall first show that Integration and Differentiation are in a certain sense, inverse operations and then define the primitive of a function and go on to prove a theorem which is usually called the Fundamental theorem of Calculus. The basic relation between the process of integration and differentiation of functions is known as Fundamental theorem of Integral calculus.

We have two Mean Value Theorems for derivatives. In the same way we have two Mean Value Theorems for integrals as well.

Theorem 8.1. *If a function f is bounded and integrable on $[a, b]$, then the function F defined as*

$$F(x) = \int_a^x f(t)dt, \quad a \le x \le b,$$

is continuous on $[a, b]$, and further more, if f is continuous at a point c of $[a, b]$, then F is derivable at c and

$$F'(c) = f(c).$$

Proof. Since f is bounded, therefore there exists a number K such that

$$|f(x)| < K, \text{ for all } x \in [a, b]$$

If x_1 and x_2 are two points of $[a, b]$ such that $a \le x_1 < x_2 \le b$, then

$$|F(x_2) - F(x_1)| = |\int_{x_1}^{x_2} f(t)dt| \le K(x_2 - x_1)$$

(using Inequality for integrals $(3)[(**)]$, chapter(6))

Thus for a given $\epsilon > 0$, we see that

$$|F(x_2) - F(x_1)| < \epsilon, \text{ if } |x_2 - x_1| < \tfrac{\epsilon}{K}.$$

Hence the function f is continuous (in fact uniformly continuous) on $[a, b]$.

Let f be continuous at a point c of $[a, b]$. Therefore, for any $\epsilon > 0$, there exists $\delta > 0$ such that

$$|f(x) - f(c)| < \epsilon, \text{ for } |x - c| < \delta$$

Let $c - \delta < s \le c \le t < c + \delta$. Then

$$|\tfrac{F(t)-F(s)}{t-c} - f(c)| = |\tfrac{1}{t-s} \int_s^t \{f(x) - f(c)\}dx| \le \tfrac{1}{t-s} \int_s^t |f(x) - f(c)|dx < \epsilon.$$

This implies that $F'(c) = f(c)$,

i.e. continuity of f at any point of $[a, b]$ implies derivability of F at that point.   □

NOTE 1:

i) From above theorem it follows that continuity of f on $[a, b]$ implies derivability of F on $[a, b]$.

ii). Above theorem is sometimes referred to as the first fundamental theorem of Integral Calculus.

Definition 8.2. (Primitive of a function): Let $f : [a, b] \to \mathbb{R}$ be a function. A primitive of f, also called indefinite Integral of f, is defined as a derivable function F, if it exists such that its derivative F' is equal to given function f i.e.

$$F'(x) = f(x) \text{ for all } x \in [a, b]$$

for example, $sinx$ is the primitive of $cosx$ over $[a, b] \subset \mathbb{R}$.

Remark 8.3. The above theorem shows that a sufficient condition for a function to admit of a primitive is that it is continuous. Thus every continuous function f possess a primitive F where

$$F(x) = \int_a^b f(t)dt$$

But the continuity of a function is not necessary condition for existence of a primitive or we can say that "functions possessing primitives are not necessarily continuous".

For example, consider the function f on $[0, 1]$ defined as

$$f(x) = 2xsin(\tfrac{1}{x}) - cos(\tfrac{1}{x}), \text{ if } x \neq 0$$
$$f(x) = 0, \text{ if } x = 0$$

It has a primitive F, where

$$F(x) = x^2 sin\tfrac{1}{x}, \text{if } x \neq 0$$
$$F(x) = 0, \text{ if } x = 0$$

Clearly $F'(x) = f(x)$ but $f(x)$ is not continuous at $x = 0$, i.e. $f(x)$ is not continuous on $[0, 1]$

Theorem 8.4. *(Fundamental Theorem of Calculus):*

If a function f is bounded and integrable on $[a,b]$ and there exists a function F such that $F' = f$ on $[a,b]$, then

$$\int_a^b f dx = F(b) - F(a).$$

Proof. Since the function $F' = f$ is bounded and integrable, therefore, for every given $\epsilon > 0$ there exists $\delta > 0$ such that for every partition

$$P = \{a = x_0, x_1, ..., x_n = b\}, \text{ with norm } \mu(P) < \delta,$$

we have

$$|\sum_{i=1}^{n} f(t_i)\Delta x_i - \int_a^b f dx| < \epsilon$$

or

$$lim_{\mu(P) \to 0} \sum_{i=1}^{n} f(t_i)\Delta x_i = \int_a^b f dx, \tag{223}$$

for every choice of point t_i in Δx_i.

Since we have freedom in the selection of points t_i in Δx_i, we choose them in particular way as follows:

By Lagrange's Mean value theorem, we have

$$F(x_i) - F(x_{i-1}) = F'(t_i) = F'(t_i)\Delta x_i \ (i = 1, 2, ..., n)$$

$$\text{or } F(x_i) - F(x_{i-1}) = f(t_i)\Delta x_i \ (i = 1, 2, ..., n)$$

i.e. $\sum_{i=1}^{n} f(t_i)\Delta x_i = \sum_{i=1}^{n}\{F(x_i) - F(x_{i-1})\} = F(b) - F(a)$.

Therefore, from (223), we have

$$\int_a^b f dx = F(b) - F(a).$$

It is sometimes referred to as *The Second Fundamental Theorem of Integral Calculus.*   □

Exercise 8.5. Show that $\int_0^1 f dx = \frac{2}{3}$, where the function f is integrable and defined as:

$$f(x) = \frac{1}{2^n}, \text{ when } \frac{1}{2^{n+1}} < x \le \frac{1}{2^n}, (n = 1, 2, ...,)$$

$$f(0) = 0.$$

Solution: The function f is integrable

Therefore, $\int_{\frac{1}{2^n}}^1 f dx = \int_{\frac{1}{2}}^1 f dx + \int_{\frac{1}{2^2}}^{\frac{1}{2}} f dx + \int_{\frac{1}{2^3}}^{\frac{1}{2^2}} f dx + ... + \int_{\frac{1}{2^n}}^{\frac{1}{2^{n-1}}} f dx$

i.e. $\int_{\frac{1}{2^n}}^1 f dx = \int_{\frac{1}{2}}^1 dx + \frac{1}{2}\int_{\frac{1}{2^2}}^{\frac{1}{2}} dx + \frac{1}{2^2}\int_{\frac{1}{2^3}}^{\frac{1}{2^2}} dx + ... + \frac{1}{2^{n-1}}\int_{\frac{1}{2^n}}^{\frac{1}{2^{n-1}}} dx$

i.e. $\int_{\frac{1}{2^n}}^1 f dx = \frac{1}{2} + \frac{1}{2}(\frac{1}{2} - \frac{1}{2^2}) + \frac{1}{2^2}(\frac{1}{2^2} - \frac{1}{2^3}) + ... + \frac{1}{2^{n-1}}(\frac{1}{2^{n-1}} - \frac{1}{2^n})$

i.e. $\int_{\frac{1}{2^n}}^1 f dx = \frac{1}{2}\{1 + \frac{1}{2^2} + (\frac{1}{2^2})^2 + (\frac{1}{2^2})^3 + ... + (\frac{1}{2^2})^{n-1}\}$

So $\int_{\frac{1}{2^n}}^1 f dx = \frac{1}{2} \cdot \frac{1 - (\frac{1}{4})^n}{1 - \frac{1}{4}} = \frac{2}{3}(1 - \frac{1}{4^n})$.

Proceeding to limits when $n \to \infty$, we get

$$\int_0^1 f dx = \frac{2}{3}.$$

Exercise 8.6. If f is a non-negative continuous function on $[a, b]$ and $\int_a^b f dx = 0$, then prove that $f(x) = 0$ for all $x \in [a, b]$.

Solution: Let $P = \{a = x_0, x_1, ..., x_n = b\}$ be a partition of $[a, b]$. Since f is continuous, therefore, f is integrable on $[a, b]$. Now

$$lim_{\mu(P) \to 0} S(P, f) = \int_a^b f dx = 0$$

Therefore,

$$0 = lim_{\mu(P) \to 0} S(P, f) = \sum_{i=1}^n f(t_i) \Delta x_i, t_i \in \Delta x_i$$

Also each term on the right side exists and is non negative and moreover for all partitions P with $\mu(P) \to 0$ and for all positions of t_i in Δx_i, we have

$$\sum_{i=1}^{n} f(t_i)\Delta x_i = 0$$

i.e. $f(t_i) = 0, i = 1, 2, 3, ...$

or, $f(x) = 0$, for all $x \in [a, b]$.

Exercise 8.7. Show that $\int_0^t sinx \; dx = 1 - cost$.

Solution: The function $sinx$ is bounded and continuous in any interval
$[0, t]$ and is therefore integrable.

Now consider a partition

$$P = \{0, \frac{t}{n}, 2\frac{t}{n}, ..., n\frac{t}{n}\} \text{ of } [0, t].$$

Now

$$U(P, sinx) = \frac{t}{n}\{sin\frac{t}{n} + sin\frac{2t}{n} + sin\frac{3t}{n} + ... + sin\frac{nt}{n}\}$$

or, $U(P, sinx) = \frac{t}{n}(cos\frac{t}{2n} + cos(n+1)\frac{t}{n})/(2sin\frac{t}{2n})$

or, $U(P, sinx) = \dfrac{\{cos\frac{t}{2n} - cos(n+1)\frac{t}{n}\}}{(\frac{sin\frac{t}{2n}}{\frac{t}{2n}})}$

and $L(P, sinx) = \frac{t}{n}\{sin\frac{t}{n} + sin\frac{2t}{n} + sin\frac{3t}{n} + ... + sin\overline{n-1}\frac{t}{n}\}$

or, $L(P, sinx) = \dfrac{\{cos\frac{t}{2n} - cos\frac{nt}{n}\}}{(\frac{sin\frac{t}{2n}}{\frac{t}{2n}})}$

In the limit,

$$U(P, sinx) = L(P, sinx) = 1 - cost$$

Therefore, $\int_0^t sinx \; dx = 1 - cost$

Theorem 8.8. *(First Mean Value Theorem):*

If a function f is continuous on $[a, b]$, then there exists a number ξ in $[a, b]$ such that

$$\int_a^b f \, dx = f(\xi)(b - a)$$

Proof. Since f is continuous, therefore f is integrable on $[a, b]$.

Let m and M be the infimum and supremum of f respectively in $[a, b]$. Then as in (178), we have

$$m(b - a) \leq \int_a^b f \, dx \leq M(b - a)$$

Therefore, there is a number $\mu \in [m, M]$, such that

$$\int_a^b f \, dx = \mu(b - a)$$

Since f is continuous on $[a, b]$, it attains every value between its bounds m, M. Therefore, there exists a number $\xi \in [a, b]$ such that $f(\xi) = \mu$.

$$\text{i.e. } \int_a^b f \, dx = f(\xi)(b - a)$$

$\square$

Remark 8.9. The above Theorem (8.8) says that the condition of continuity is necessary for the function to assume its mean value in the given interval, for example the function $f(x)$ defined on $[2, 5]$ as follows:

$f(x) = 1$ if $2 \leq x < 3$;

$f(x) = 3$ if $3 \leq x \leq 5$.

Now

$$\int_2^5 f \, dx = \int_2^3 1 \, dx + \int_3^5 3 \, dx = 7$$

So the mean value of the function is

$$\tfrac{1}{5-2} \int_2^5 f \, dx = \tfrac{7}{3}$$

which the function fails to assume this value on the interval.

Theorem 8.10. *(Generalized first Mean Value Theorem):*

If f and g are integrable on $[a,b]$ and g keeps the same sign over $[a,b]$, then there exists a number μ lying between the bounds of f such that

$$\int_a^b fg \, dx = \mu \int_a^b g \, dx.$$

Proof. Let g be a positive over $[a,b]$.

If m and M are the bounds of f, then for all $x \in [a,b]$, we have

$$m \leq f(x) \leq M$$

i.e. $mg(x) \leq f(x)g(x) \leq Mg(x)$

Therefore, for $b \geq a$

$$m \int_a^b g(x) \, dx \leq \int_a^b f(x)g(x) \, dx \leq M \int_a^b g(x) \, dx \qquad (224)$$

Let μ be the number lying between m and M.

Then,

$$\int_a^b fg \, dx = \mu \int_a^b g \, dx \qquad (225)$$

$\square$

Corollary 8.11. *If in addition to the conditions of above theorem, f is* *continuous on* $[a, b]$*, there exists a number* $\xi \in [a, b]$*, such that*

$$\int_a^b fg\,dx = f(\xi) \int_a^b g\,dx \tag{226}$$

NOTE 2: i) If $g(x) \le 0$, the sign of inequality changes in (224) but (225) and (226) remain unchanged.

ii) If $b \le a$, the sign of inequality changes in (224) but (225) and (226) remain unchanged.

Theorem 8.12. *Integration by Parts:*

If f and g are integrable on $[a, b]$ with their primitives $F(x)$ and $G(x)$, and

$$F(x) = A + \int_a^x f\,dx, \quad G(x) = B + \int_a^x g(x)\,dx$$

where A and B are constants, then

$$\int_a^b F(x)g(x)\,dx = [F(x)G(x)]_a^b - \int_a^b G(x)f(x)\,dx$$

where $[F(x)G(x)]_a^b$ denotes the difference $F(b)G(b) - F(b)G(a)$.

Proof. Let $P = \{a = x_0, x_1, ..., x_n = b$ be a partition of $[a, b]$.

We have

$$[F(x)G(x)]_a^b = \sum_{i=1}^n [F(x_i)G(x_i) - F(x_{i-1})G(x_{i-1})]$$

$$= \sum_{i=1}^n G(x_i)[F(x_i) - F(x_{i-1})] + \sum_{i=1}^n F(x_{i-1})[G(x_i) - G(x_{i-1})]$$

or,

$$[F(x)G(x)]_a^b = G(x_i) \int_{x_{i-1}}^{x_i} f(x)dx + \sum_{i=1}^{n} F(x_{i-1}) \int_{x_{i-1}}^{x_i} g(x)dx \qquad (227)$$

Let $\Delta f_i = f(x_i) - f(x_{i-1})$ and $\Delta g_i = g(x_i) - g(x_{i-1})$ be the oscillation of f and g in Δx_i.

Now for all $x \in \Delta x_i$, we have

$$|f(x) - f(x_i)| \leq |f(x_i) - f(x_{i-1})| = \Delta f_i \text{ and}$$

$$|g(x) - g(x_{i-1})| \leq \Delta g_i$$

This implies that $f(x_i) - \Delta f_i \leq f(x_i) + \Delta f_i$ and

$$g(x_i) - \Delta g_i \leq g(x) \leq g(x_{i-1}) + \Delta g_i$$

i.e. $[f(x_i) - \Delta f_i]\Delta x_i \leq \int_{x_{i-1}}^{x_i} f(x)dx \leq [f(x_i) + \Delta f_i]\Delta x_i$ and

$$[g(x_{i-1}) - \Delta g_i]\Delta x_i \leq \int_{x_{i-1}}^{x_i} g(x)dx \leq [g(x_{i-1}) + \Delta g_i]\Delta x_i$$

Therefore,

$$\int_{x_{i-1}}^{x_i} f(x)dx = [f(x_i) + \theta_i \Delta f_i]\Delta x_i \qquad (228)$$

$$\text{and} \int_{x_{i-1}}^{x_i} g(x)dx = [g(x_{i-1}) + \theta_i' \Delta g_i]\Delta x_i \qquad (229)$$

where $-1 \leq \theta_i, \theta_i' \leq 1$.

Hence from (227), (228) and (229), we get

$$[F(x)G(x)]_a^b = \sum G(x_i)f(x_i)\Delta x_i + \sum F(x_{i-1})g(x_{i-1})\Delta x_i + \sigma \qquad (230)$$

where $\sigma = \sum [G(x_i)\Delta f_i \theta_i + F(x_{i-1})\Delta g_i \theta_i']\Delta x_i$

Now F and G, being continuous, are bounded, therefore, a number k exists such that

$|F(x)| \leq k, |G(x) \leq k|$, for all $x \in [a, b]$

Therefore, $|\sigma| \leq k(\sum \Delta f_i + \sum \Delta g_i)\Delta x_i$

In the limit when $\mu(P) \to 0$, $\sigma \to 0$, and therefore, (230) implies that

$$[F(x)G(x)]_a^b = \int_a^b G(x)f(x)dx + \int_a^b F(x)g(x)dx$$

or

$$\int_a^b F(x)g(x)dx = [F(x)G(x)]_a^b - \int_a^b G(x)f(x)dx$$

Hence the proof. $\qquad\qquad\square$

Corollary 8.13. *If a function g is bounded and integrable on $[a, b]$ and if a function f is derivable and its derivative f' is bounded and integrable on $[a, b]$, then*

$$\int_a^b f(x)g(x)dx = [f(x)\int_a^x g(x)dx]_a^b - \int_a^b \{f'(x)\int_a^x g(x)dx\}dx$$

or, $\int_a^b f(x)g(x)dx = f(b)\int_a^b g(x)dx - \int_a^b \{f'(x)\int_a^x g(x)dx\}dx$

If however both the derivatives f' and g' are assumed to be bounded and integrable, a much shorter and simpler proof exists, which follows.

Theorem 8.14. *(A particular case). If f and g are both differentiable on $[a, b]$ and if f' and g' are both integrable on $[a, b]$ then*

$$\int_a^b f(x)g'(x)dx = [f(x)g(x)]_a^b - \int_a^b g(x)f'(x).$$

Proof. Since f and g are differentiable and hence continuous on $[a, b]$, therefore f and g are integrable over $[a, b]$. Again, since f, g, f' and g' are all integrable over $[a, b]$. therefore, $f\,g'$, $g\,f'$ are integrable over $[a, b]$.

Let $F(x) = f(x)g(x)$, for all $x \in [a, b]$,

Now $F'(x) = f(x)g'(x) + g(x)f'(x)$

Therefore, $\int_a^b F'(x)dx = \int_a^b \{f(x)g'(x) + g(x)f'(x)\}dx$

$$\text{or} \int_a^b F'(x)dx = \int_a^b \{f(x)g'(x) + \int_a^b g(x)f'(x)\}dx \qquad (231)$$

Also by Fundamental Theorem (**??**),

$$\int_a^b F'(x)dx = F(b) - F(a) = [f(x)g(x)]_a^b \qquad (232)$$

Now from (231) and (232), we get

$$\int_a^b \{f(x)g'(x) = [f(x)g(x)]_a^b - \int_a^b g(x)f'(x)dx \qquad \qquad \square$$

Theorem 8.15. *(Change of the variable in an integral):*

If i) f is integrable over $[a,b]$,

ii) ϕ is a derivable, strictly monotonic function on $[\alpha, \beta]$, where $a = \phi(\alpha)$, $b = \phi(\beta)$, and

iii) g' is integrable over $[\alpha, \beta]$, then

$$\int_a^b f(x)dx = \int_\alpha^\beta f(\phi(y))\phi'(y)dy$$

[change of variable in $\int_a^b f(x)dx$ by putting $x = \phi(y)$]

Proof. Let ϕ be strictly monotonic increasing on $[\alpha, \beta]$.

Since ϕ is strictly monotonic, it is invertible, i.e.

$x = \phi(y)$ implies $y = \phi^{-1}(x)$, for all $x \in [a, b]$. Therefore,

$\alpha = \phi^{-1}(a)$ and $\beta = \phi^{-1}(b)$

Let $P = \{a = x_0, x_1, x_2, ..., x_n = b\}$ be any partition of $[a, b]$ and

$Q = \{\alpha = y_0, y_1, y_2, ..., y_n = \beta\} = \phi^{-1}(x_i)$ the corresponding partition of $[\alpha, \beta]$.

By Lagrange's Mean Value Theorem (4.11),

$$\Delta x_i = \phi(y_i) - \phi(y_{i-1}) = \phi'(\eta_i)\Delta y_i, \eta_i \in \Delta y_i \qquad (233)$$

Let

$$\xi = \phi(\eta_i), \; where \; \xi \in \Delta x_i \qquad (234)$$

Now

$$S(P, f) = \sum_{i=1}^{n} f(\xi_i)\Delta x_i = \sum_{i=1}^{n} f(\phi(\eta_i))\phi'(\eta_i)\Delta y_i \qquad (235)$$

Uniform continuity of ϕ implies that $\mu(Q) \to 0$ as $\mu(P) \to 0$

Also f is integrable implies $lim_{\mu(p)\to 0}S(P, f)$ exists and therefore, the limit of the term on right hand side of (235) also exists as $\mu(Q) \to 0$ and this limit equals the integral $\int_\alpha^\beta f(\phi(y))\phi'(y)dy$.

Hence letting $\mu(P) \to 0$ in (235), we get

$$\int_a^b f(x)dx = \int_\alpha^\beta f(\phi(y))\phi'(y)dy$$

$\square$

Remark 8.16. If $\phi'(y) \neq 0$ for any $y \in [\alpha, \beta]$, then ϕ is strictly monotonic in $[\alpha, \beta]$. Hence the condition of monotonicity of ϕ in the statement of the above Theorem (8.15), may be replaced by

$$\phi'(y) \neq 0$$

for all $y \in [\alpha, \beta]$.

Remark 8.17. Theorem 8.15 holds even if $\phi'(y) = 0$ for a finite number of values of y. For, in that case, the the interval $[\alpha, \beta]$ can be divided into a finite number of sub-intervals, in each of which ϕ is strictly monotonic. The repetition of the argument for each of sub-intervals in turn and addition of the results, gives the required result.

Note: The proof of Second Mean Value Theorem depends on Abel's Lemma. First of all we will prove the Lemma.

Lemma 8.18. *(Abel's Lemma):*

If $\{b_n\}$ is a positive monotonic decreasing sequence and k, K denote respectively the least and the greatest values of the sums $\sum_{r=m}^{p} u_r$, for $p = m, m+1, ..., n$, then

$$b_m k \leq \sum_{r=m}^{n} b_r u_r \leq b_m k$$

Proof. Let $S_P = \sum_{r=m}^{p} u(r)$. Then

$\sum_{r=m}^{p} b_r u(r) = b_m u_m + b_{m+1} u_{m+1} + ... + b_n u_n$, which implies that

$\sum_{r=m}^{p} b_r u(r) = b_m S_m + b_{m+1}(S_{m+1} - S_m) + ... + b_n(S_n - S_{n-1})$

i.e. $\sum_{r=m}^{p} b_r u(r)$

$= (b_m - b_{n+1})S_m + (b_{m+1} - b_{m+2})S_{m+1} + ... + (b_{n-1} - b_n)S_{n-1} + b_n S_n$

Now the terms (within brackets) on the right are non-negative, therefore,

$k(b_m - b_{m+1} + b_{m+1} - b_{m+2} + ... + b_{n-1} - b_n + b_n)$

$\leq \sum_{r=m}^{n} b_r u_r$

$\leq K(b_m - b_{m+1} + ... - b_n + b_n)$

Therefore, $b_m k \leq \sum_{r=m}^{n} b_r u_r \leq b_m K$

In particular for $m = 1$, the Lemma may be stated in the form:

If $b_1, b_2, ..., b_n$ is a positive monotonic decreasing set and k, K denote respectively the least and the greatest values of the partial sums, $\sum_{r=1}^{p} u_r$, $1 \leq p \leq n$ of the members, $u_1, u_2, ..., u_n$ then

$$b_1 k \leq \sum_{r=1}^{n} b_r u_r \leq b_1 K.$$

$\square$

Theorem 8.19. *(Second Mean Value Theorem):*

If $\int_a^b f dx$ and $\int_a^b g dx$ both exist and f is monotonic on $[a,b]$, then there exists $\xi \in [a,b]$ such that

$$\int_a^b fg \ dx = f(a) \int_a^{\xi} g \ dx + f(b) \int_{\xi}^b g \ dx$$

Proof. We first prove the Bonnett's form of the Theorem, where in addition to the hypothesis of the theorem, the monotone function is positive and monotone decreasing on $[a,b]$, i.e Bonnett's theorem states that:

If $\int_a^b \phi \ dx$ and $\int_a^b g \ dx$ both exists and ϕ is positive and monotone decreasing on $[a,b]$, then there exists a point $\xi \in [a,b]$ such that

$$\int_a^b \phi g \ dx = \phi(a) \int_a^{\xi} g \ dx$$

Let $P = \{a = x_0, x_1, ..., x_n = b\}$ be a partition of $[a,b]$, and M_t, m_t the bounds of g in Δx_i.

Let $t_1 = a$ and $t_i (in \neq 1)$ be any point of Δx_i.

In the sub-interval Δx_i, we know

$$m_i \Delta x_i \leq \int_{x_{i-1}}^{x_i} g \ dx \leq M_i \Delta x_i$$

and

$$m_i \Delta x_i \leq g(t_i) \Delta x_i \leq M_i \Delta x_i$$

Letting $i = 1, 2, 3, ..., p, \ (p \leq n)$ and adding, we get

$$\sum_{i=1}^{p} m_i \Delta x_i \leq \int_a^{xp} g \ dx \leq \sum_{i=1}^{p} M_i \Delta x_i$$

and

$$\sum_{i=1}^{p} m_i \Delta x_i \leq \sum_{i=1}^{p} g(t_i) \Delta x_i \leq \sum_{i=1}^{p} M_i \Delta x_i$$

which gives

$$\left| \int_{p}^{xp} g dx - \sum_{i=1}^{n} g(t_i) \Delta x_i \right| \leq \sum_{i=1}^{p} (M_i - m_i) \Delta x_i \leq w(P, g)$$

$$\text{i.e.} \int_{a}^{xp} g dx - w(P, g) \leq \sum_{i=1}^{p} g(t_i) \Delta x_i \leq \int_{a}^{xp} g dx + w(P, g),$$

where $w(P, g)$ denotes the oscillatory sum, $U(P, g) - L(P, g)$.

Now, $\int_{a}^{t} g \, dx$, being a continuous function is bounded. Let A, B be its bounds. Then we have

$$B - w(P, g) \leq \sum_{i=1}^{p} g(t_i) \Delta x_i \leq A + w(P, g) \qquad (236)$$

Using Abel's Lemma (8.18), where

$b_t = \phi(t_1), u_1 = g(t_1) \Delta x_i$

$k = B - w(P, g), K = A + w(P, g);$

we get

$$\phi(a)[B - w(P, g)] \leq \sum_{i=1}^{n} \phi(t_i) g(t_i) \Delta x_i \leq \phi(a)[A + w(P, g)]$$

Taking the limit when $\mu(P) \to 0$, we get

$$B\phi(a) \leq \int_{a}^{b} \phi g \, dx \leq A\phi(a)$$

$$\Rightarrow \int_{a}^{b} \phi g \, dx = \mu\phi(a) \qquad (237)$$

where μ is some number between B and A.

The function $\int_a^b g(x)dx$ being continuous, must assume, for some $\xi \in [a,b]$, the value μ, which lies between its bounds. Thus we get

$$\int_a^b \phi g \, dx = \phi(a) \int_a^\xi g \, dx \tag{238}$$

Proof of the Main Theorem:

Let first, f be monotone decreasing, so that the function ϕ where $\phi f - f(b)$ is positive and monotone decreasing. Then, by what has been proved above, there exists a number ξ between a and b such that

$$\int_a^b g[f - f(b)]dx = [f(a) - f(b)] \int_a^\xi g dx$$

$$i.e. \int_a^b fg dx = f(a) \int_a^\xi g \, dx + f(b)[\int_a^b g dx - \int_a^\xi g \, dx]$$

or,

$$\int_a^b fg \, dx = f(a) \int_a^\xi g dx + f(b) \int_\xi^b g \, dx \tag{239}$$

Let now f be monotone increasing, so that $(-f)$ is monotonic decreasing and therefore, from (238),

$$\int_a^b (-f)g \, dx = -f(a) \int_a^\xi g \, dx - f(b) \int_\xi^b g \, dx$$

$$i.e. \int_a^b fg \, dx = f(a) \int_a^\xi g \, dx + f(b) \int_\xi^b g \, dx$$

Hence the Theorem. $\square$

Note: It may be easily verified that the theorem holds for $a > b$ also.

Theorem 8.20. *(Second Mean Value Theorem, a particular case):*

If f is monotonic and f, f' and g are all continuous in $[a,b]$, then there exists $\xi \in [a,b]$ such that

$$\int_a^b f(x)g(x) \; dx = f(a) \int_a^\xi g(x) \; dx + f(b) \int_\xi^b g(x) \; dx$$

Proof. Let $G(x) = \int_a^x g(t) \; dt$. Now clearly $G(a) = 0$, and under the given conditions, $G(x)$ is differentiable and $G'(x) = g(x)$.

$$\text{Therefore, } \int_a^b f(x)g(x)dx = \int_a^b f(x)G'(x)dx =$$
$$[f(x)G(x)]_a^b - \int_a^b G(x)f'(x)dx].$$

Since G being continuous is integrable and f is monotonic and continuous on $[a, b]$, therefore, on using generalized First Mean Value Theorem (8.10), there exists $\xi \in [a, b]$ such that

$$\int_a^b f(x)g(x)dx = f(b)G(b) - G(\xi) \int_a^b f'(x)dx$$

$$\text{i.e. } \int_a^b f(x)g(x)dx = f(b)G(b) - G(\xi)\{f(b) - f(a)\}$$

$$\text{i.e. } \int_a^b f(x)g(x)dx = f(b)\{G(b) - G(\xi)\} + f(a)G(\xi)$$

$$\text{i.e. } \int_a^b f(x)g(x)dx = f(b) \int_a^b g(x)dx + f(a) \int_a^\xi g(x)dx$$

$$\square$$

Lemma 8.21. *(Riemann-Lebesgue Lemma):*

If a function f is bounded and integrable on $[a, b]$, show that

$$lim_{n \to \infty} \int_a^b f(x)cosnx \; dx = 0 \;\; and \;\; lim_{n \to \infty} \int_a^b f(x) \; dx = 0$$

Proof. Let $I_n = \int_a^b f(x)cosnnx \; dx$.

Let, further, $\epsilon > 0$ be an arbitrary number.

Now f is bounded and integrable on $[a, b]$, there exists a partition $P = \{a = x_0, x_1, x_2, ..., x_p = b\}$ such that the oscillatory sum

$$U(P, f) - L(P, f) = \sum_{i=1}^{p}(M_i - m_i)\Delta x_i < \epsilon/2$$

where M_i, m_i are bounds of f in Δx_i. Now

$$I_n = \sum_{i=1}^{p} \int_{x_{i-1}}^{x_i} f(x)cosnx \, dx$$

i.e. $I_n = \sum_{i=1}^{p} f(x_{i-1}) \int_{x_{i-1}}^{x_i} cosnx \, dx +$
$\sum_{i=1}^{p} \int_{x_{i-1}}^{x_i} \{f(x) - f(x_{i-1})\}cosnx \, dx$

Therefore,

$$|I_n| \leq \sum_{i=1}^{p} |f(x_{i-1})| |\int_{x_{i-1}}^{x_i} cosnx \, dx| +$$
$$\sum_{i=1}^{p} \int_{x_{i-1}}^{x_i} |\{f(x) - f(x_{i-1})\}cosnx| \, dx$$

But for all $x \in \Delta x_i$, we have

$$|f(x) - f(x_{i-1})| \leq M_i - m_i, \, (i = 1, 2, ..., p)$$

Therefore,

$$|\{f(x) - f(x_{i-1})\} \, cosnx| \leq M_i - m_i$$

and

$$\sum_{i=1}^{n} \int_{x_{i-1}}^{x_i} |\{f(x)f(x_{i-1})\}cosnx| \, dx \leq \sum_{i=1}^{p}(M_i - m_i)(x_i - x_{i-1}) < \tfrac{1}{2}\epsilon$$

Also $|\int_{x_{i-1}}^{x_i} cosnx \, dx| = |\tfrac{1}{n}\{sin \, nx_1 - sin \, nx_{i-1}\}|$

i.e. $|\int_{x_{i-1}}^{x_i} cosnx \, dx| \leq \tfrac{1}{n}\{|sin \, nx_1| + |sinn \, x_{i-1}|\} \leq \tfrac{2}{n}$

or, $|I_n| \leq \tfrac{2}{n} \sum_{i=1}^{p} |f(x_{i-1})| + \tfrac{1}{2}\epsilon$

For a fixed P, $f(x_{i-1})$ is a fixed quantity. Also there exists a positive number m such that for all $n \geq m$,

$$\tfrac{2}{n} \sum_{i=1}^{p} |f(x_{i-1})| < \tfrac{1}{2}\epsilon$$

Therefore, $|I_n| < \epsilon$, for all $n \geq m$

i.e. $lim_{n\to\infty} \int_a^b f(x)cosnx \ dx = 0$.

It may be shown that

$$lim_{n\to\infty} \int_a^b f(x)sinnx \ dx = 0$$

$\square$

Note: In particular, if $f(x)$ is bounded and integrable in $[0, \frac{1}{2}\pi]$, then

$$lim_{n\to\infty} \int_0^{\frac{\pi}{2}} f(x)dx = 0$$

If we assign the value 0 at the origin to $(\frac{1}{x} - \frac{1}{sinx})$, it becomes continuous, bounded and integrable in $[0, \frac{1}{2}\pi]$ and so

$$lim_{n\to\infty} \int_0^{\frac{\pi}{2}} f(x)(\frac{1}{x} - \frac{1}{sinx}))sin \ nx \ dx = 0$$

i.e. $lim_{n\to\infty} \int_0^{\frac{\pi}{2}} f(x)\frac{sinnx}{x} dx = lim_{n\to\infty} \int_0^{\frac{\pi}{2}} f(x)\frac{sinnx}{sinx} dx$

Exercise 8.22. Use the First Mean Value theorem of Integral calculus (8.8) to show that

$$\frac{\pi}{6} < \int_0^{\frac{1}{2}} \frac{dx}{\sqrt{[(1-x^2)(1-k^2x^2)]}} \leq \frac{\pi}{6} \frac{1}{\sqrt{1-k^2/4}}$$

where $k^2 < 1$

Solution: Let $f(x) = \frac{1}{\sqrt{1-x^2}}$, $g(x) = \frac{1}{\sqrt{1-k^2x^2}}$, we see $g(x) > 0$ on $[0, \frac{1}{2}]$.

Also $1 < g(\xi) < \frac{1}{\sqrt{\frac{k^2}{4}}}$ for any $x \in [0, \frac{1}{2}]$ and $k^2 > 1$.

Now by First Mean Value Theorem (8.8), we get

$$\int_0^{1/2} \frac{dx}{\sqrt{(1-x^2)(1-k^2x^2)}} = g(\xi)\int_0^{1/2}\frac{dx}{\sqrt{1-x^2}}$$

Hence

$$\int_0^{1/2}\frac{dx}{\sqrt{1-x^2}} = g(\xi)\int_0^{1/2}\frac{dx}{\sqrt{1-x^2}} < \frac{1}{\sqrt{\frac{k^2}{4}}}\int_0^{1/2}\frac{dx}{\sqrt{1-x^2}}$$

as $f(x) \geq 0$

$$Hence,\ \frac{\pi}{6} < \int_0^{\frac{1}{2}}\frac{dx}{\sqrt{[(1-x^2)(1-k^2x^2)]}} \leq \frac{\pi}{6}\frac{1}{\sqrt{1-\frac{k^2}{4}}}$$

Exercise 8.23. If a function f is continuous on $[0,1]$, then show that

$$lim_{n\to\infty}\int_0^1 \frac{nf(x)}{1+n^2x^2}dx = \frac{\pi}{2}f(0)$$

Solution: We have

$$\int_0^1 \frac{nf(x)}{1+n^2x^2}dx = \int_0^{\frac{1}{\sqrt{n}}}\frac{nf(x)}{1+n^2x^2}dx + \int_{\frac{1}{\sqrt{n}}}^1 \frac{nf(x)}{1+n^2x^2}dx$$

By generalized first Mean Value Theorem (8.10), we have

$$\int_0^{\frac{1}{\sqrt{n}}}\frac{nf(x)}{1+n^2x^2}dx = f(\xi)\int_0^{\frac{1}{\sqrt{n}}}\frac{n}{1+n^2x^2}dx,\ \text{where}\ 0 \leq \xi \leq \frac{1}{\sqrt{n}}$$

$$\text{or,}\ \int_0^{\frac{1}{\sqrt{n}}}\frac{nf(x)}{1+n^2x^2}dx = f(\xi)tan^{-1}\sqrt{n} \to \frac{\pi}{2}f(0)\ \text{as}\ n \to \infty$$

Again since f is continuous on $[0,1]$, it is bounded and therefore, there exists K such that

$$|f(x)| \leq K,\ \text{for all}\ x \in [0,1].$$

Therefore,

$$\left| \int_{\frac{1}{\sqrt{n}}}^{1} \frac{nf(x)}{1+n^2x^2} dx \right|$$

$$\leq K \left| \int_{\frac{1}{\sqrt{n}}}^{1} \frac{n}{1+n^2x^2} dx \right|$$

$$= K \left| tan^{-1}n - tan_{-1}\sqrt{n} \right|$$

$$\to 0 \text{ as } n \to \infty.$$

Hence the result.

EXERCISES

(1) If f is continuous and non-negative on $[a, b]$ and $\int_a^b f dx = 0$, then prove that $f(x) = 0$, for all $x \in [a, b]$.

(2) If $f : [0, 1] \to \mathbb{R}$ is defined as

$f(x) = x$; if x is rational

$f(x) = 0$; if x is irrational.

Discuss the Riemann integrability of f.

(3) If $f(x) = 0$ or 1 according as x is rational or non rational, prove that f is not integrable on any interval.

(4) Prove that the function f defined as

$f(x) = x$, when x is rational

$f(x) = -x$, when x is not rational

is not integrable over $[a, b]$; but $|f|$ is integrable.

(5) If f is continuous on $[a, b]$ and $\int_a^b f^2 dx = 0$, then prove that $f \equiv 0$ on $[a, b]$.

(6) If f, g are integrable on [a, b] then prove that

$$\int_a^b f^2 dx + \int_a^b g^2 dx \leq 2 \int_a^b |fg| dx$$

(7) Integrate on $[0, 2]$ the function $f(x) = x[x]$, where $[x]$ denotes the greatest integer not greater than x.

(8) Evaluate $\int_0^2 f(x)\, dx$

where $f(x) = 0$, when $x = \frac{n}{n+1}, \frac{n+1}{n}$, n= 1, 2, 3,...

$f(x) = 1$, elsewhere.

Is f integrable on $[0, 1]$?

(9) Test the applicability of second Mean Value Theorem to the functions:

(a) $f(x) = \cos x$ and $g(x) = x^2$

(b) $f(x) = e^n$, $g(x) = x$

(10) If $f \in R[a, b]$, then prove that for $\epsilon > 0$, there exists $g \in C[a, b]$ such that $\int_a^b |f - g| dx < M$.

(11) If $f : [-1, 1] \to \mathbb{R}$ is defined as

$f(x) = x \sin \frac{1}{x^2} - \frac{1}{x} \cos \frac{1}{x^2}$, when $x \neq 0$

$f(x) = 0$, when $x = 0$.

Then prove that f has a primitive but it is not Riemann integrable.

(12) if $f : [a, b] \to \mathbb{R}$ is Riemann integrable and $\frac{1}{f}$ is bounded on $[a, b]$, then prove that $\frac{1}{f}$ is Riemann integrable on $[a, b]$.

(13) Show hat the function f defined on $[0, 1]$ as

$f(x) = \sqrt{1 - x^2}$ if x is rational

$f(x) = 1 - x$ if x is irrational.

is not Riemann integrable.

(14) If f is bounded on (a, b) and is Riemann integrable on every closed subinterval of (a, b), then show that f is integrable on (a, b).

(15) If f and g are integrable and

$$\int_a^b f dx = \int_a^b g dx,$$

then show that

$f(\xi) = g(\xi)$, for some $\xi \in [a, b]$.

(16) Distinguish between Integral and Primitive of a function.

(17) If $f(x) = \frac{1}{n}$, for $\frac{1}{n+1} < x \leq \frac{1}{n}$, $n = 1, 2, 3, \ldots$ and $f(0)$ has any assigned value, prove that $f(x \in R[0, 1]$ and

$$\int_0^1 f(x) dx = \frac{\pi^2}{6} - 1.$$

SUBJECT INDEX